INTRODUCTORY MICROBIOLOGY LAB SKILLS AND TECHNIQUES IN FOOD SCIENCE

INTRODUCTORY MICROBIOLOGY LAB SKILLS AND TECHNIQUES IN FOOD SCIENCE

CANGLIANG SHEN

Food Microbiologist and Associate Professor/Extension Specialist, Division of Animal and Nutritional Sciences, Davis College of Agriculture, Natural Resources and Design, West Virginia University, Morgantown, WV, USA

YIFAN ZHANG

Professor, Department of Nutrition and Food Science, Wayne State University, Detroit, MI, USA

Academic Press is an imprint of Elsevier
125 London Wall, London EC2Y 5AS, United Kingdom
525 B Street, Suite 1650, San Diego, CA 92101, United States
50 Hampshire Street, 5th Floor, Cambridge, MA 02139, United States
The Boulevard, Langford Lane, Kidlington, Oxford OX5 1GB, United Kingdom

Notices

Knowledge and best practice in this field are constantly changing. As new research and experience broaden our understanding, changes in research methods, professional practices, or medical treatment may become necessary.

Practitioners and researchers must always rely on their own experience and knowledge in evaluating and using any information, methods, compounds, or experiments described herein. In using such information or methods they should be mindful of their own safety and the safety of others, including parties for whom they have a professional responsibility.

To the fullest extent of the law, neither the Publisher nor the authors, contributors, or editors, assume any liability for any injury and/or damage to persons or property as a matter of products liability, negligence or otherwise, or from any use or operation of any methods, products, instructions, or ideas contained in the material herein.

British Library Cataloguing-in-Publication Data

A catalogue record for this book is available from the British Library

Library of Congress Cataloging-in-Publication Data

A catalog record for this book is available from the Library of Congress

ISBN: 978-0-12-821678-1

Publisher: Susan Dennis
Acquisitions Editor: Nina Bandeira
Editorial Project Manager: Regine A. Gandullas
Production Project Manager: Kumar Anbazhagan
Cover Designer: Victoria Pearson

Typeset by Aptara, New Delhi, India

Dedication

I dedicate this book to Dr. Peter Schaeffer, Interim Division Director of Animal and Nutritional Sciences, Professor of Resource Economics and Management, West Virginia University. Dr. Schaeffer retired from West Virginia University on July 31, 2021 after serving for more than 20 years.

Contents

Preface

I have been teaching food microbiology lab course for food science and human nutrition graduate and undergraduate students for 6 years at West Virginia University. I found that more than 60% of the students lack the basic microbial lab skills when conducting food microbiology lab course tasks. For example, lots of them failed to get single colonies by streak-plating technique, and at least half of the students cannot operate bright-field microscope independently. We know that students who take food microbiology lab course should take 200 or 300 level general microbial lab course as prerequisite class, however many small institutions may not offer this type of enteral level general microbial lab classes, especially with the budget cut of high education these days. Therefore, I authored this general microbiology lab book which is specific for food science and human nutrition students. This book could be used as a supplement material for food microbiology lab and other undergraduate level microbial lab courses. The topics include in this book are isolation, identification, numeration and observation of microorganisms, biochemistry tests, case studies, clinical lab tasks, and basic applied microbiology. The book is written technically with figures showing each lab task steps by steps, and the pictures are taken directly from students' action in real lab classroom.

Acknowledgments

I appreciate my previous and current division directors Drs. Robert Taylor, Kim Barnes, and Peter Schaeffer. They all supported me to develop and teach general microbiology and food microbiology laboratory courses since I joined West Virginia University in 2015. I appreciate my PhD advisor Dr. John N. Sofos, University Distinguished Professor of Colorado State University for his mission of "safety, quality, and quantity" as the three principles to conduct any microbiology related lab works. I would love to thank my PhD student Dr. KaWang Li, who assisted to take pictures of all chapters in this book. Finally, I would love to thank my wife Dr. Yi Xu and my 4-year-old son Alex Shen for their love and support.

-Cangliang Shen

CHAPTER 1

Introduction of microbial and chemical safety and lab notebook record

Chapter outline

1.1 Lab safety rules

The following basic rules/regulations must be followed at all times in order to promote and maintain a high level of safety in the laboratory.

Personal conduct in the laboratory should always be courteous and professional. Personal attention in the principles of safety are always in order.

Because most microbiological laboratory procedures require the use of living organisms, an integral part of all laboratory sessions is the use of aseptic technique. Although the virulence of microorganisms used in the laboratory has been greatly diminished, all microorganisms should be treated as potential pathogens (organisms capable of causing disease). Thus, students must carefully adhere to aseptic techniques as described by the instructor. It is important for you to develop a positive and respectful attitude toward microorganisms and your work with them. Casualness in technique can cause needless contamination of your work and infection of yourself and others.

Do not hesitate to ask questions if a laboratory procedure is unclear.

Before entering the laboratory, place coats, books, and other paraphernalia in the locker that is assigned to you. Be aware that these lockers are not locked. Valuables can be stored in a specified location in the laboratory as directed by the laboratory instructor.

Never trust the student who worked at your laboratory bench before you, but always respect the student who comes after you. At the beginning and termination of each laboratory session, wipe down bench tops with disinfectant solution provided by the instructor.

DO NOT EAT, DRINK, OR SMOKE in the laboratory. Make every effort to minimize hand-to-mouth contact. Lick-on labels are prohibited

Introductory Microbiology Lab Skills and Techniques in Food Science
DOI: https://doi.org/10.1016/B978-0-12-821678-1.00021-6

from use in the laboratory. Never apply cosmetics or insert contact lenses in the laboratory.

Lab coats/aprons, protective safety glasses and disposable exam gloves must be worn while working in the laboratory. Safety glasses will be provided (box on the lab bench). Some aprons are available for use during the semester. These are available on a first-come, first-serve basis. If you have or will have other laboratory courses that require a lab coat, you may wish to purchase you own. Each student will be required to purchase exam gloves. Failure to wear these items will result in dismissal from the laboratory. Do not wear loose clothing (e.g., scarves) that could catch fire or knock over reagents/cultures. **Do not wear open-toed shoes** such as sandals in the laboratory.

Never remove media, equipment, or especially bacterial cultures from the laboratory unless given specific permission from the laboratory instructor.

Long hair must be secured in place to prevent injury from Bunsen burner flames.

Wash your hands with liquid soap provided and dry with paper towels upon entering and prior to leaving the laboratory.

Keep only those materials on your desktop that you need to perform your experiment. Workspace is at a minimum; cluttering it with unneeded items such as books will increase the risk of a laboratory accident.

Do not place contaminated instruments, such as pipettes, on bench top. Inoculating loops and needles must be sterilized by incineration (as described by laboratory instructor) and pipettes must be disposed of in designated receptacles.

On completion of the laboratory session, place all cultures and materials in the disposal area as designated by the laboratory instructor (laboratory cart at front of laboratory). **Never discard contaminated liquids or liquid cultures into the sink.** Put them into designated containers in the lab for later sterilization.

Carry cultures in appropriate containers when moving the laboratory. Likewise, keep cultures in a test tube rack on bench tops when not in use. This serves a dual purpose to prevent accidents and avoid contamination of yourself and the environment.

Uncontaminated materials like the lens paper and paper towels must be placed in wastebaskets. Discard all paper and plastic materials as soon as practical. There is always the danger of fire if there is an accumulation of combustibles.

If liquid cultures are spilled, immediately cover with paper towels and then saturate with the disinfectant solution. Notify the laboratory instructor who will dispose of contaminated materials appropriately.

Report all accidental injuries (cuts, burns, etc.) to the instructor **immediately**. If you have an open wound before coming to the laboratory, notify the instructor at the start of the laboratory period. You may be prohibited from working with live cultures that day.

Never pipette by mouth any broth cultures or chemical reagents. Doing so is strictly prohibited. Pipetting is to be carried out with the aid of a mechanical pipetting device.

Be certain that you are aware of the location and familiar with the use of emergency safety items (eye wash station, safety shower, and first aid supplies) located in the laboratory. Use these items without delay since time is an extremely important factor in minimizing further injury. Familiarize yourself with the location of the fire alarm and emergency exit route (in case of fire).

In the event of an **active shooter,** extinguish Bunsen burners, turn out lights to the laboratory and quietly and quickly make your way through one of the two interior doors in the laboratory.

Small bottles of alcohol are located on the bench top. **Keep Bunsen burners away from alcohol bottles** since a flame can ignite the alcohol and cause serious injury.

Some of the chemical reagents used in the laboratory are considered hazardous by the Federal The Environmental Protection Agency (EPA). You will be given specific instructions by the laboratory instructor on the handling of these materials at the appropriate time in the semester. In most cases, the laboratory instructor will assume responsibility for handling and disposing of these potentially hazardous chemicals. If you have questions about the potential hazards or proper handling of any chemicals, you have the right-to-know and information will be provided. Material Safety Data Sheets of hazardous chemicals are available in the laboratory for your use.

If you are a person with a disability/limitation concern and anticipate needing any type of accommodation in order to participate in this class, contact Disability Services (304-293-6700) to determine what action is appropriate. Please Note: if you suspect that are immunocompromised (diabetic, immune deficiency disease, pregnant, etc.) it is your responsibility to contact your physician to determine whether additional safety precautions are needed in order for you to safely complete the laboratory requirement for this course.

Lab notebook record: Notebooks can be checked after each lab section before you leave the classroom. The front cover of your notebook will be labeled with following information: Your name, course name, professor's name, semester. Reserve the first 3 pages of your notebook for a table of contents. You will update the table of contents each week. Number all pages. Each notebook entry for any given lab period will be formatted in the following way:

1. The Title.
2. A Purpose statement: Describe the purpose of the exercise. Ask yourself "What are the goals of this exercise?"
3. A Materials and Methods section - The Materials and Methods for each lab will described at the beginning of the period by the instructor. You will have an opportunity to copy this information into your notebook at the beginning of each lab period. You should describe what was needed and the steps taken (including any modifications that were made). Be sure to use CORRECT SPELLING for all microorganism names, and italicize scientific names. Points will be deducted for incorrectly formatted scientific names. NOTE: Copying from the lab manual is illegal.
4. A Results section: Record all observations in your lab notebook. Colored pencils/pens should be used to illustrate results (i.e., observations made with the microscope) - all figures/tables must have a title and legend (a description of what is being shown - label all relevant information).
5. A Discussion - Summarize your findings and discuss how the exercise helped you understand the learning objectives. Describe why something may not have worked, and what you would do differently next time to improve the outcome. Each section must be LABELED. Points will be deducted for incomplete entries and sloppiness.

CHAPTER 2

Practice of bright-field microscope

Chapter outline

2.1 Materials

Bright-field microscope, prepared slides of *Staphylococcus aureus* and *Candida albicans*

2.2 Introduction

Microorganism are defined as live bodies are too small to see using naked eyes, therefore microorganisms are typically observed under microscope. Microscopes are categorized as light and electronic microscope. Light microscopes include bright-field, dark-field, phase-contrast, fluorescence, and laser confocal microscope. Bright-field microscope is the most common tool in the microbiology lab to observe microorganisms, especially for bacteria, yeast, and molds.

Structure of bright-field microscope is shown in Fig. 2.1, a bright-field microscope composed by two most important lens, including ocular lens and objective lens with holding by body tubes and adjusted by the course focus (think knob) and find focus (thin knob). The microbial specimens are prepared on glass slides and placed on the center of the stage with the stage control moving back and forward. A light source controlled by the power switch on the sides, and a condenser are located underneath the stage with an iris diaphragm to control the amount of lights passing through the specimen.

Introductory Microbiology Lab Skills and Techniques in Food Science
DOI: https://doi.org/10.1016/B978-0-12-821678-1.00011-3

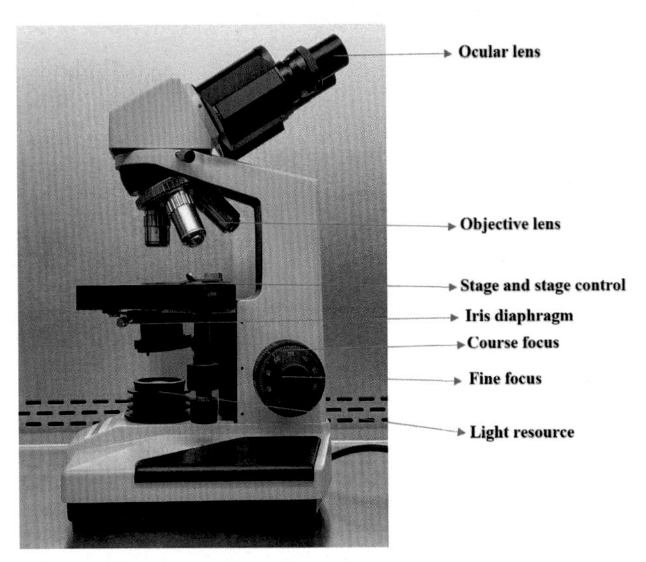

Fig. 2.1 ***Structure of bright-field microscope.***

2.3 Terminologies of bright-field microscope

1. Total magnification: It is calculated by the magnification of ocular lens multiple by the magnification of objective lens. The magnification of ocular lens is always set at 10X. For example, calculation of the total magnification of a microscope using a 10X ocular with a 100X objective is (10X) × (100X) = 1000X.
2. Working distance: It is the distance between objective lens to the stage of specimen. The working distance decreases with the increase of the magnification of objective lens.
3. Parfocal: Once the microscope has been focused, it stays focus during switching between objective lens. For example, if the microscope was focus under 10X, then it is switched to 40X, or to 100X, only needs little adjustment with fine focus knob.
4. Oil immersion: Light passing through various substances causes bending referred as refraction, which is increased when the objective lens switches from 10X, to 40X, and to 100X. The refraction is measured by the refractive index (R.I.) with the $R.I._{air} = 1.0$ and $R.I._{glass} = 1.5$. The excessive refraction causes the distortion of the images. Since RI value of oil is 1.5 equals that of glass ($R.I._{oil} = R.I._{glass} = 1.5$, adding one large drop of oil onto the center of the microscope slide can remove the air gap between specimen and objective lens, usually this is practiced with 100X objective lens.

5. Resolution power (RP): It is the capability to differentiate the two small parts underneath microscopes. It is calculated as RP = λ/2 × N.A. λ equals 650 nm if voltage control set at "10" or at largest. N.A. refers as numerical aperture ("size of the engineer"), as shown in Fig. 2.2, under 100X objective lens the N.A. is 1.25. Therefore, the RP of a bright-field microscope is calculated as RP = 650 nm/2 × 1.25 = 260 nm = 0.26 μm.

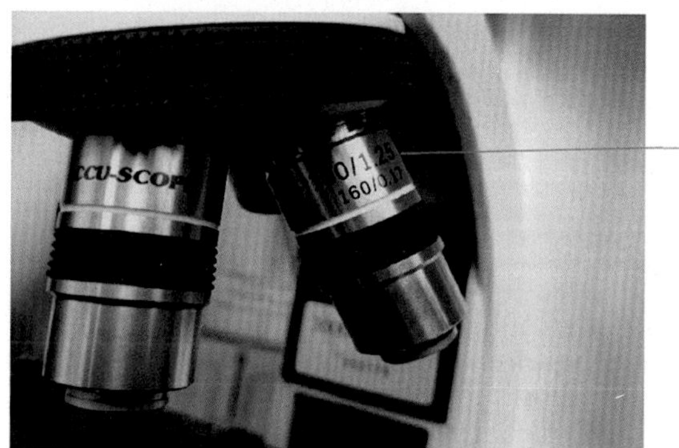

Figure of 100X objective lens

Fig. 2.2 *100X objective lens showing numeric aperture.*

2.4 Procedure of focusing prepared slides

1. Obtain a prepared slide from the front desk
2. Insert the slide into the stage of the microscope as shown in Fig. 2.3

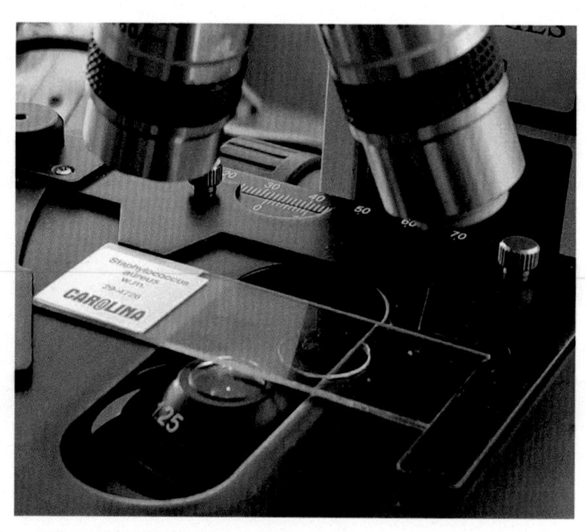

Fig. 2.3 *Put the slide onto the stage center.*

3. Bring the stage down to the lowest position using course focus Fig. 2.4

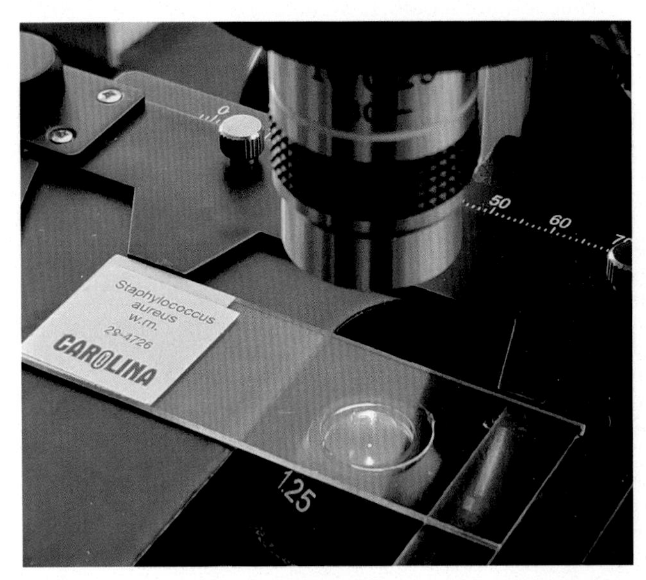

Fig. 2.4 ***Stage down and switch 10X objective lens to the center Move the stage to the highest position.***

4. Switch 10X objective lens to the center
5. Move the stage with the slides to the highest position
6. Make adjustment using fine focus, set iris diagram at "10" as shown in Fig. 2.5

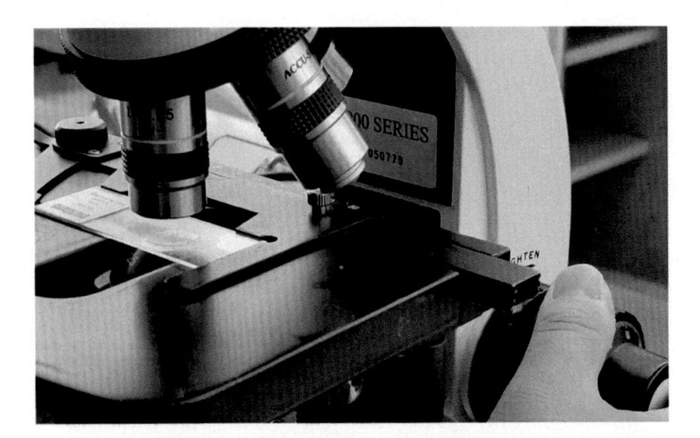

Fig. 2.5 ***Make adjustment using course focus, set iris diagram at "10."***

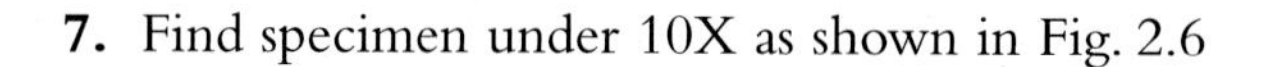

7. Find specimen under 10X as shown in Fig. 2.6

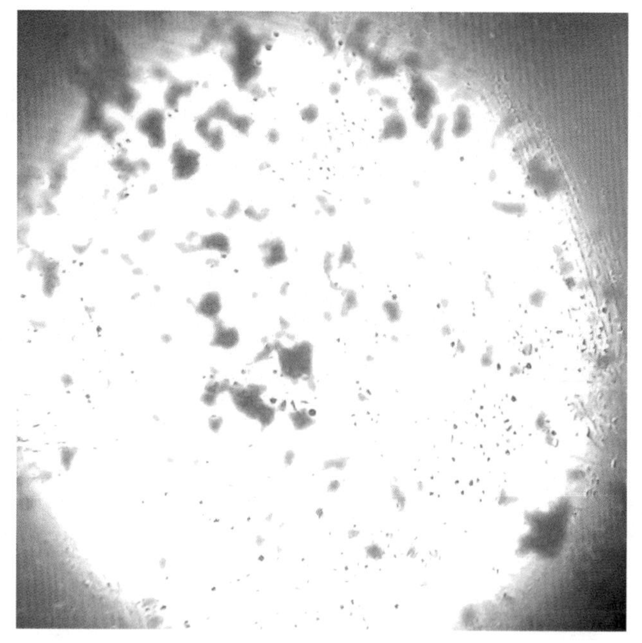

Fig. 2.6 ***Find specimen under 10X objective lens.***

8. Slightly switch objective lens to make space for adding one large drop of oil as shown in Fig. 2.7
9. Switch 100X objective lens to the center

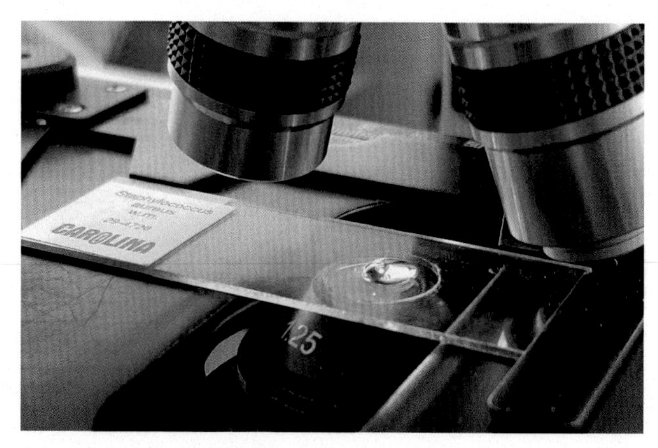

Fig. 2.7 ***Switch objective lens and add one drop of oil.***

10. Make adjustment using fine focus, set iris diagram at "100" as shown in Fig. 2.8

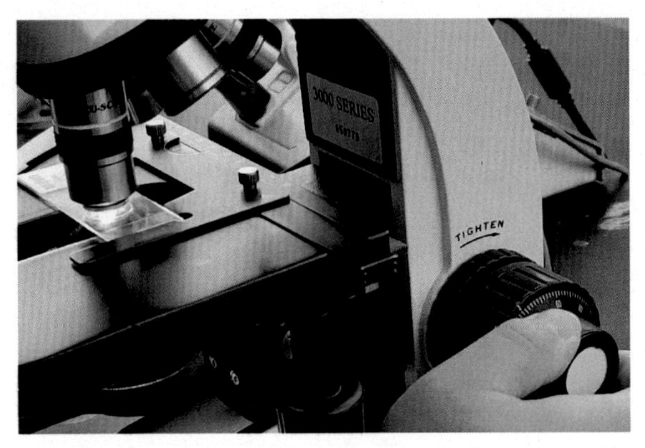

Fig. 2.8 ***Switch to 100X objective lens using fine focus for slight adjustment, set iris diagram at "100."***

11. Draw the morphology of the specimen on your notebook or taking pictures as shown in Fig. 2.9

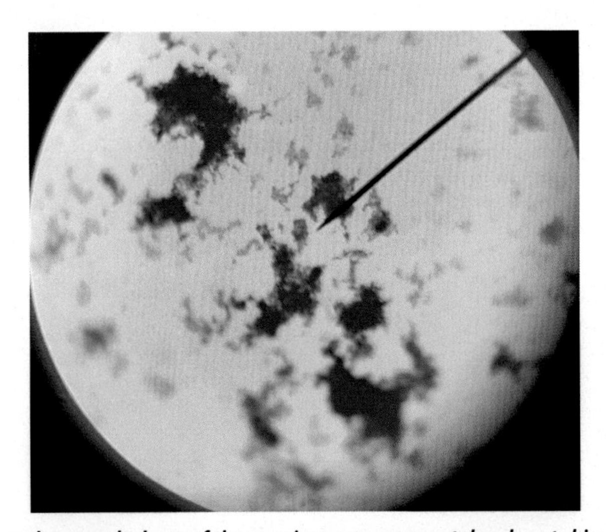

Fig. 2.9 ***Draw the morphology of the specimen on your notebook or taking pictures.***

12. Remove oil from the slides using lens paper
13. Clean ocular and objective lens using lens paper
14. Place back the microscope to the cabinet with the black cover

2.5 After class question

How would a decrease in the wavelength of light affect the resolution power?

CHAPTER 3

Preparation of smears and simple stain practice

Chapter outline

3.1 Materials

Bright-field microscope, glass slides, crystal violet, safranin, malachite green, methylene blue, carbol fuchsin, *Escherichia coli*, *Staphylococcus aureus*, *Bacillus cereus*, *Micrococcus luteus* (on agars or on slants)

3.2 Introduction

Bacteria needs to be stopped of their motion and stick themselves with something in order to be observable to our naked eyes. The easiest way to achieve this is to prepare a smear of bacterial cells, fix them onto a glass slide and simple stain with a chemical dye. Basic dyes, including crystal violet, safranin, malachite green, and methylene Blue, has a positively charged chromophore that is attracted to the negatively charged bacterial cell outer surfaces. Underneath bright field microscope with the magnification of 100X with oil immersion, the shape of bacterial cells, referred as morphology, can be clearly seen.

3.3 Morphology of bacterial cells in this lab practice

Escherichia coli: short, single, rod shape
Bacillus cereus: large, single, rod, chain shape
Staphylococcus aureus: round, cocci, clustered in grape shape
Micrococcus luteus: round, cocci, clustered in tetrads shape

Introductory Microbiology Lab Skills and Techniques in Food Science
DOI: https://doi.org/10.1016/B978-0-12-821678-1.00014-9

3.4 Procedure

1. Obtain a glass slide and wipe it with a drop of 95% alcohol using a bench paper towel.
2. Label 1 circle as (E.C; B.C.; S.A.; or M.L.) on the bottom of your slides with a marker as shown in Fig. 3.1.

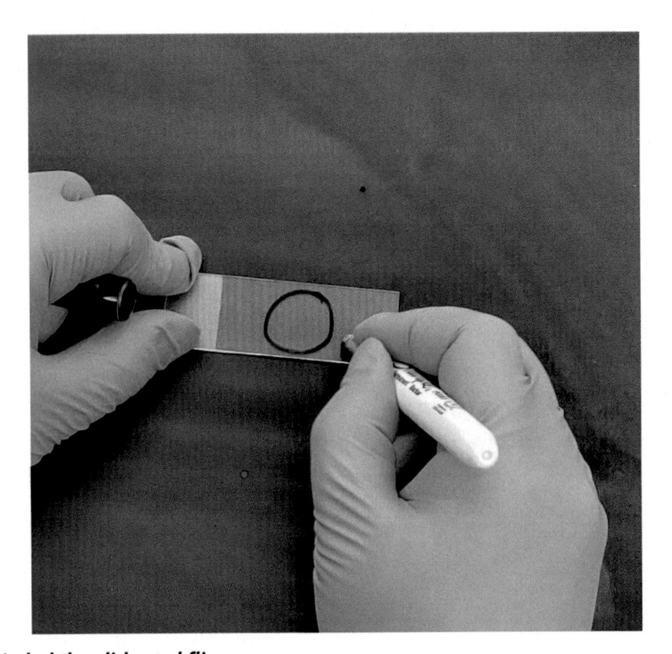

Fig. 3.1 ***Label the slide and flip over.***

3. Flip your marked glass slides, otherwise the sharpie marker may interfere with the dye or the lens of microscope.
4. Flame the loop as shown in Fig. 3.2 and add one drop of water into the center of the circle as shown in Fig. 3.3.
5. Aseptically pick cells from agars or slants Fig. 3.4 and spread them with the drop of water into the circles as shown in Fig. 3.5.
6. Flame the loop or change with new loop before and after each transfer to prevent cross-contamination.
7. All the smear to completely air dry for 3–5 min.

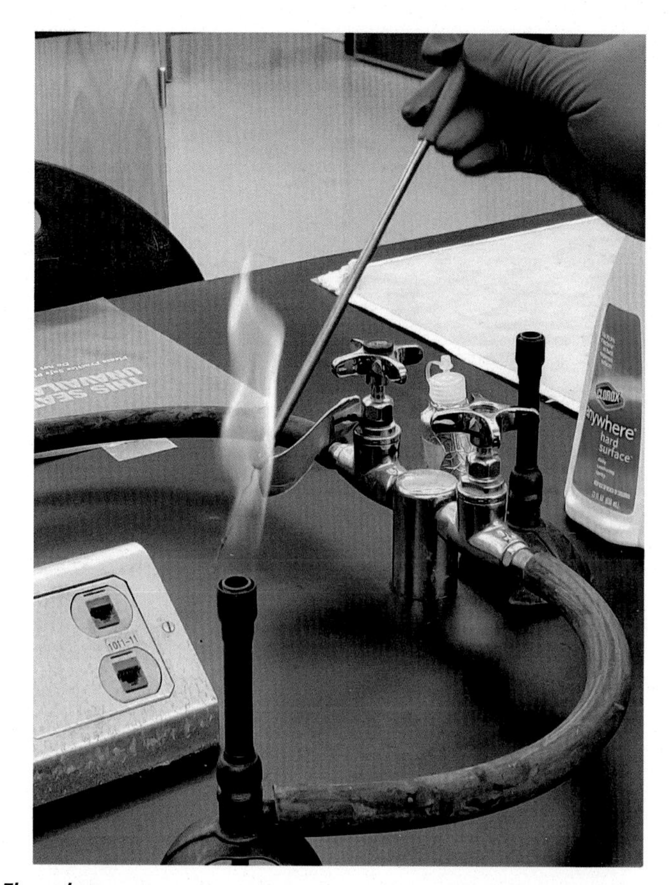

Fig. 3.2 ***Flame loop.***

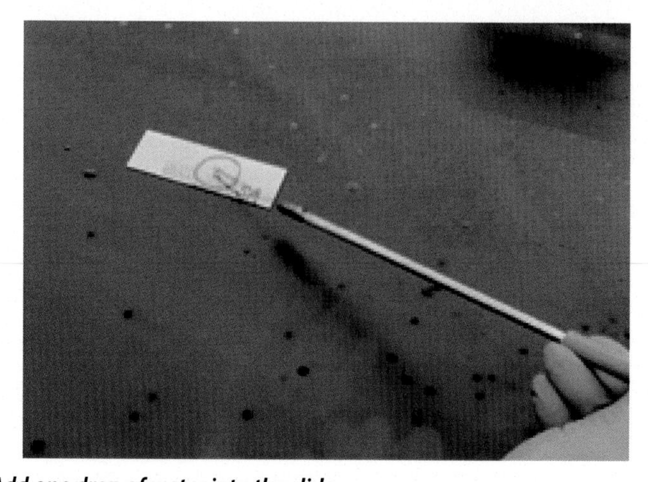

Fig. 3.3 ***Add one drop of water into the slide.***

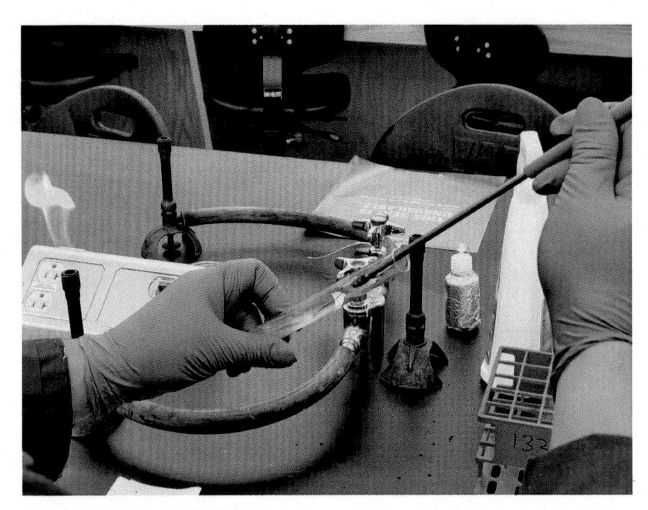

Fig. 3.4 ***Pick bacteria from the slant.***

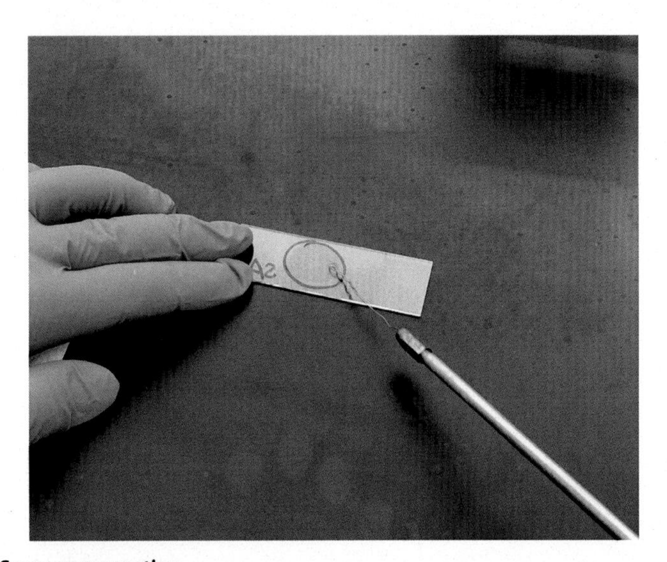

Fig. 3.5 ***Smear preparation.***

8. Heat fixing the smear onto the slide by passing the glides through the flamed Bunsen Burner 3-5 times, which will assist stick bacterial cells onto the slides by inactivate their enzymes as shown in Fig. 3.6.
9. Place the slide on the staining rack and add one drop of basic dye into the center of the circle as shown in Fig. 3.7.

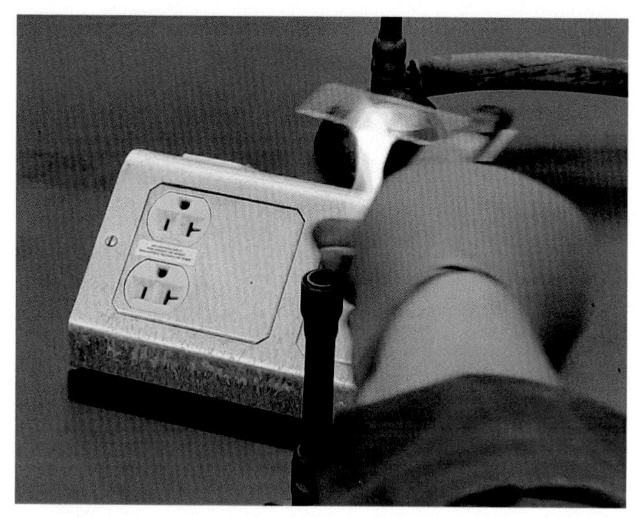

Fig. 3.6 ***Heat fixing.***

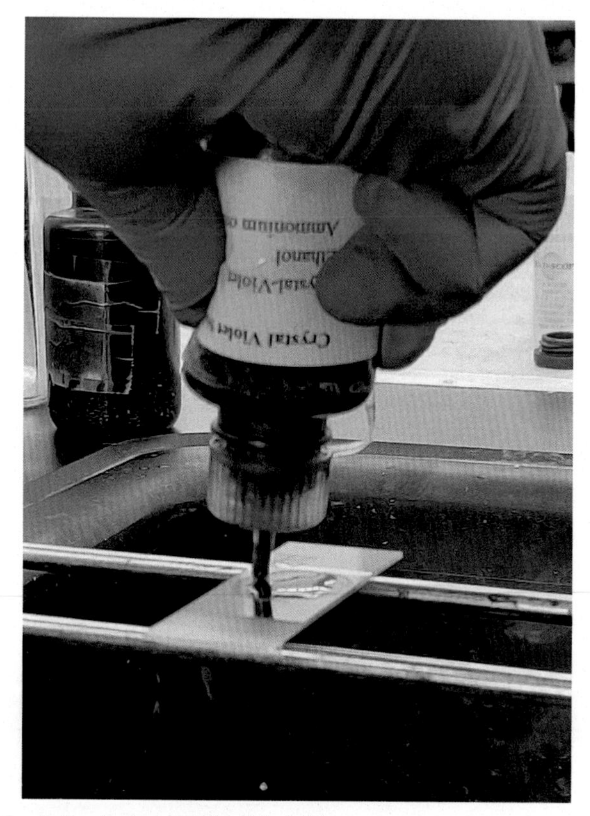

Fig. 3.7 ***Add basic dye in the middle of the slide.***

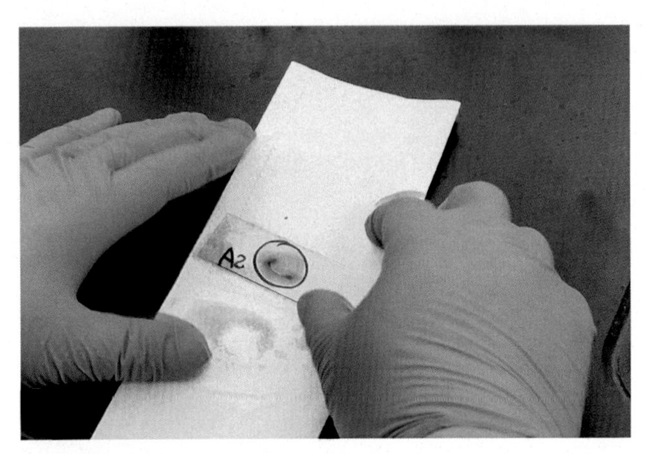

Fig. 3.8 ***Dry on paper towel***

10. After 1 min, gently remove the dye with running tap water or water in the bottles.
11. Drain off the excess water.
12. Dry with covering the slides with paper towel with gentle pressure as shown in Fig. 3.8.
13. Find the specimen under brightfield microscope with 10X objective lens.
14. Switch 100X objective lens with oil immersion.
15. Draw pictures, record observations, including morphology and arrangement in the results of Table 3.1.

Table 3.1 Results table of simple stain.

Bacteria names	Morphology and arrangement under 100X with oil immersion
Escherichia coli	
Staphylococcus aureus	
Bacillus cereus	
Micrococcus luteus	

3.5 After class questions

1. Why we need to do heat-fixing for smear preparation?
2. Why it is important to describe the cells morphology and arrangement?

CHAPTER 4

Gram-stain practice

Chapter outline

4.1 Materials

Bright-field microscope, glass slides, crystal violet, Gram iodine (I_2), Safranin, 90% alcohol, *Escherichia coli*, *Staphylococcus aureus*, *Micrococcus luteus* (on agars or on slants)

4.2 Introduction

Gram-stain was created by Hans Christian Gram (Danish microbiologist) to differentiate pneumococci rom *Klebsiella pneumonia* in 1884, which differentiate all bacteria into two groups as gram positive (blue/purple color) and gram negative (pink/red color). There are basically two theories explaining the Gram-stain differentiation, which are based on cell wall structure and lipids component. The first theory is referred as "Cell wall theory". Compared to the Gram-negative bacteria, a much thicker peptidoglycan layer is in the gram-positive bacteria, which assists alcohol, a decolorizer, to dehydrate the cell wall, and strongly maintains the crystal violet complex inside the cell wall and prevents the purple color of crystal violet washing off during decoloration. The second theory is referred as "lipids theory." There are greater amount (10%–15%) of lipids in the cell wall of Gram-negative than Gram-positive bacteria (~5%), which makes alcohol (decolorizer) wash off the crystal violet complex easily. Therefore, the Gram-negative bacterial cell wall is temporarily colorless after decoloration followed by counter stained with Safranin showing pink/red color. For Gram-stain, very young culture (~18–24 h) is required to be used and

Introductory Microbiology Lab Skills and Techniques in Food Science
DOI: https://doi.org/10.1016/B978-0-12-821678-1.00008-3

results need to be recorded, including three items: (1) Gram reaction, (2) cell morphology (shape of cells), and (3) cell arrangement.

4.3 Gram-stain results of bacterial cells in this lab practice

Escherichia coli: gram negative, short, single, rod shape
Staphylococcus aureus: gram positive, round, cocci, clustered in grape shape
Micrococcus luteus: gram positive, round, cocci, clustered in tetrads shape

4.4 Procedure

1. Clean your obtained glass slides with a drop of 95% alcohol with paper towel.
2. Label two circles as E.C and S.A. (or M.L.) on the bottom of your slides with a sharpie marker as shown in Fig. 4.1.

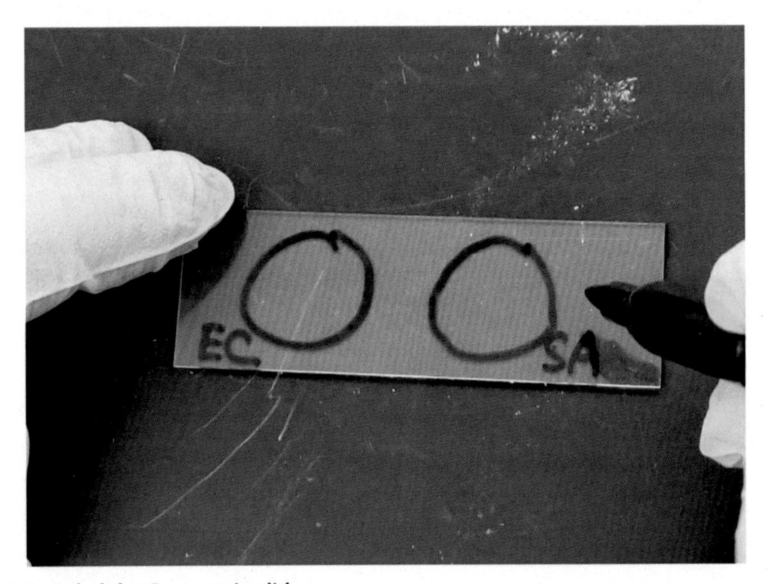

Fig. 4.1 ***Label the Gram stain slide.***

3. Flip your marked glass slides as shown in Fig. 4.2.
4. Add one drop of water into the center of the two circles.
5. Aseptically pick cells from agars or slants and spread them with the drop of water into the circles.
6. Flame the loop or change with new loop before and after each transfer to prevent cross contamination.

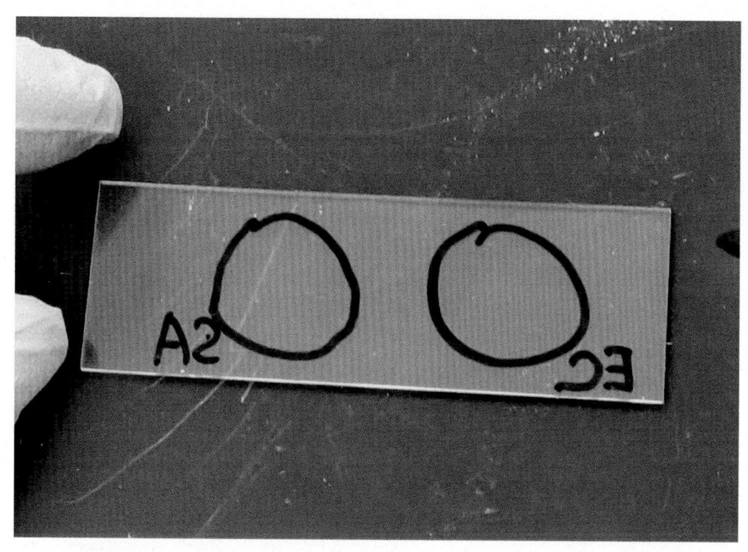

Fig. 4.2 ***Flip over the slide.***

7. All the smear to completely air dry for 3–5 min.
8. Heat fixing the smear onto the slide by passing the glides through the flamed Bunsen burner 3–5 times as shown in Fig. 4.3.

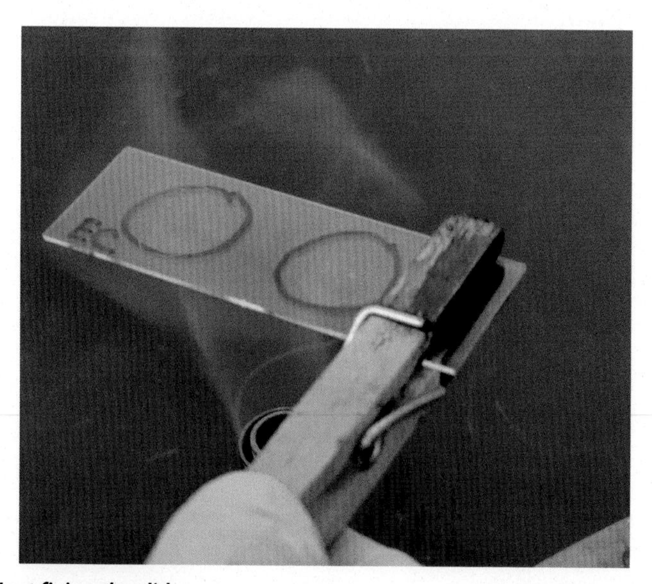

Fig. 4.3 ***Heat fixing the slide.***

9. Place the slide on the staining rack and add crystal violet into the center of the circle as shown in Fig. 4.4.

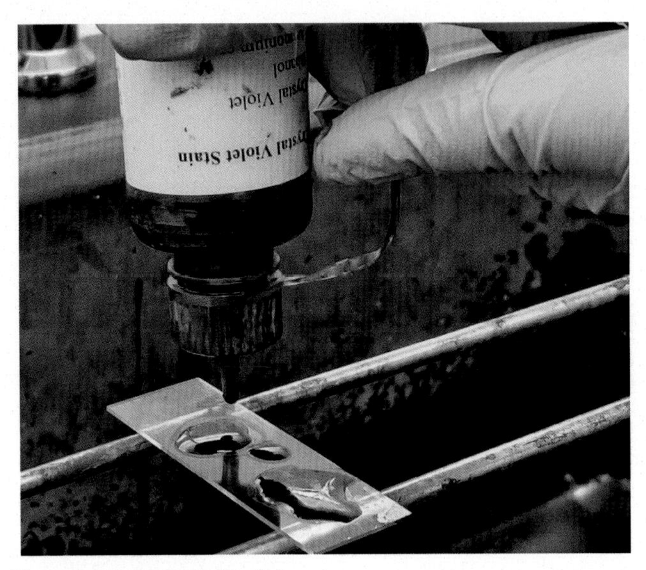

Fig. 4.4 ***Add crystal violet onto the slide.***

10. After 1 min, remove the dye with wash gently with water.
11. Add Grams iodine into the center of the circle as shown in Fig. 4.5.

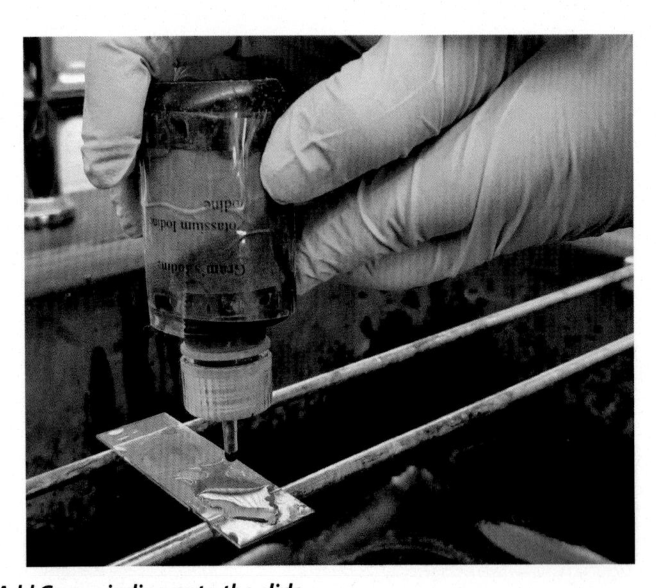

Fig. 4.5 ***Add Grams iodine onto the slide.***

12. After 1 min, remove the dye with wash gently with water.
13. Add 90% alcohol to decolor for 10 sec and rinse gently with water as shown in Fig. 4.6.

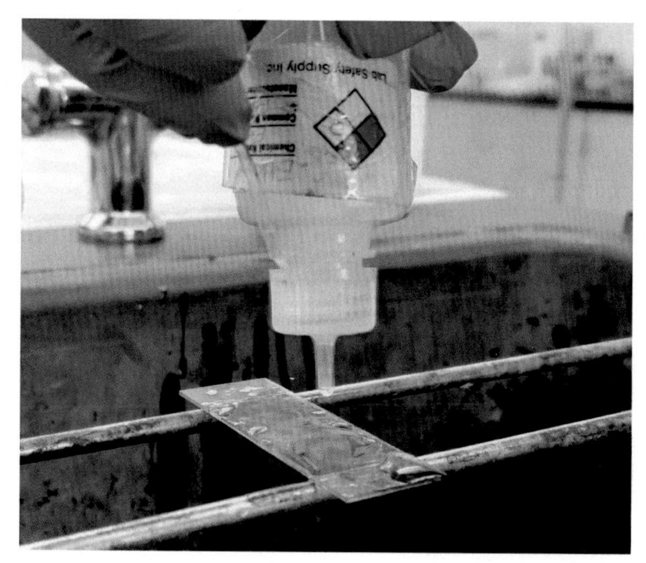

Fig. 4.6 ***Decoloration of slide with 90% alcohol.***

14. Add counter stain dye Safranin onto the center of the circle as shown in Fig. 4.7.

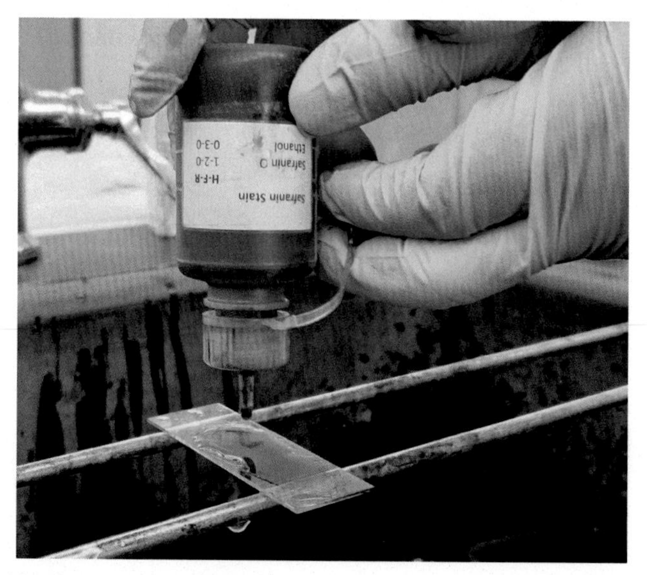

Fig. 4.7 ***Add Safranin onto the slide.***

15. After 1 min, gently remove the dye with wash gently with water.
16. Drain off excess water, blot dry with paper towel, and let the slide air dry.
17. Find the specimen under bright-field microscope with 10X objective lens.
18. Switch 100X objective lens with oil immersion.
19. Draw pictures, record observations, including Gram reaction, morphology, and arrangement in the results table (Table 4.1).

Table 4.1 Results of Gram stain.

Bacteria names	Gram reaction	Morphology and arrangement
Escherichia coli		
Staphylococcus aureus		
Micrococcus luteus		
Bacillus cereus		

4.5 After class questions

1. Describe two conditions that will cause Gram variable stain.
2. Describe the theories of Gram-stain differentiation.
3. Why Gram-stain is important for describing bacteria in clinical microbiology area?
4. What is the advantage of the Gram-stain over the simple stain?

CHAPTER 5

Acid-fast stain practice

Chapter outline

5.1 Materials

Bright-field microscope, glass slides, carbol fuchsin, methylene blue, acid alcohol, *Mycobacterium phlei* (on agars or on slants)

5.2 Introduction

Mycobacterium has two major species *Mycobacterium tuberculosis* and *M. bovis*. *M. tuberculosis* is the etiological agent of tuberculosis (TB) in humans, and the humans are the only reservoir for the bacterium. TB of the lung is spread through the air by coughing, sneezing, or talking. Crowed area and confined environment is usually the contribution factor. *M. bovis* is the etiologic agent of TB in cows and rarely in humans. It usually infected to humans by the consumption of unpasteurized milk and linked to the bone infections that led to the hunched backs. The cell wall structure of Mycobacterium contains peptidoglycan with a unique complex of unsaturated lipids chain called mycolic acid. The existing of mycolic acid form a liquid shell around the organisms and make it a very waxy cell due to the strong hydrophobic molecules. The staining of the waxy cell wall relies on heat-steaming with carbol fuchsin followed by discoloration with acid alcohol and counterstain with methylene blue.

5.3 Clinical diagnose procedure

1. Active cough for more than 2–3 weeks with chest pain, fever, fatigue, and loss of appetite.
2. Coughing up blood in sputum.

Introductory Microbiology Lab Skills and Techniques in Food Science
DOI: https://doi.org/10.1016/B978-0-12-821678-1.00004-6

3. X-ray showing cloudy lung area.
4. Purified protein derivatives skin testing showing a pale elevation of the skin 6–10 mm in diameter (BCG vaccine may cause false positive).
5. Positive acid-fast stain bacilli of the patient sputum sample (presumptive TB).
6. Harsh alkaline digestion of sputum samples or applied in a new fluorometric mycobacteria detection system to remove background flora of sputum.
7. Transfer to transport medium Middlebrook's and Lowenstein-Jensen medium to CDC reference lab for 4–6 weeks of visual confirmation.

5.4 Procedure

1. Clean your obtained glass slides with a drop of 95% alcohol with paper towel.
2. Label one circle as M. P. (*M. phlei*) on the bottom of your slides with a sharpie marker
3. Flip your marked glass slides.
4. Add one drop of water into the center of the two circles.
5. Aseptically pick cells from agars or slants and spread them with the drop of water into the circles.
6. Flame the loop or change with new loop before and after each transfer.
7. All the smear to completely air dry for 3–5 min.
8. Heat fixing the smear onto the slide by passing the glides through the flamed Bunsen burner 3–5 times.
9. Place the slide on the staining rack and flooding carbol fuchsin onto the whole slides.
10. Heat steam slides for 5 min using the flame of burner back and forth couple of times, but not let the slides dry out.
11. Add acid alcohol (3% HCl in 95% ethanol) into the center of the circle for decolorize 20 sec.
12. After 20 sec, wash gently with water.
13. Counterstain with methylene blue for 30 sec.
14. After 30 sec, wash gently with water.
15. Drain off excess water, blot dry with paper towel, and let the slide air dry.
16. Find the specimen under brightfield microscope with 10X objective lens.
17. Switch 100X objective lens with oil immersion.

18. Draw pictures, record observations, especially morphology of acid fast and nonacid fast cells.

Please fill the results of acid fast-stain as shown in Table 5.1.

Table 5.1 Results table of acid fast-stain.

Bacteria names	Morphology and arrangement under 100X with oil immersion
Mycobacterium phlei	

5.5 After class questions

1. Is the presence of acid-fast bacilli in a sputum enough evidence of TB? Why?
2. How do clinical labs culture detect the presence of Mycobacterium?

CHAPTER 6

Endospore stain practice

Chapter outline

6.1 Materials

Bright-field microscope, glass slides, malachite green, Safranin, *Escherichia coli, Bacillus cereus* (on agars or on slants)

6.2 Introduction

Two bacteria genera will generate endospore (germination) under stressful environment (poor nutrition, low humidity, desiccation, and high temperature) are Bacillus and Clostridium. The calcium-dipicolinic acid stabilizes the endospore' DNA along with small soluble DNA protein to protect the bacteria from stress environment. When the growth condition improving, the endospore will generate vegetative cells, referred as sporulation. Endospore stain is intended to find the presence/absence of endospore and the location of endospore. The location of endospore is species specific which may be in the middle of cells (central), and the end (terminal, i.e., *Clostridium sporogenes*), or between the end and the middle (subterminal, i.e., *Clostridium botulism*). Endospore stain needs to be done on old culture (>5–7 days).

6.3 Procedure

1. Clean your obtained glass slides with a drop of 95% alcohol with paper towel.
2. Label two circles as E.C. (*E. coli*) and B.S. (*B. cereus*) on the bottom of your slides with a sharpie marker.
3. Flip your marked glass slides.

Introductory Microbiology Lab Skills and Techniques in Food Science
DOI: https://doi.org/10.1016/B978-0-12-821678-1.00013-7

4. Add one drop of water into the center of the two circles.
5. Aseptically pick cells from agars or slants and spread them with the drop of water into the circles.
6. Flame the loop or change with new loop before and after each transfer.
7. All the smear to completely air dry for 3–5 min.
8. Heat fixing the smear onto the slide by passing the glides through the flamed Bunsen burner 3–5 times.
9. Place the slide on the staining rack and flooding Malachite green onto the whole slides as shown in Fig. 6.1.

Fig. 6.1 ***Flooding slide with malachite green.***

10. Heat steam slides for 5 min using the flame of burner back and forth couple of times as shown in Fig. 6.2, but not let the slides dry out.
11. Add Safranin (Fig. 6.3) into the center of the circle as shown in Fig. 6.4.
12. After 1 min, gently remove the dye with wash gently with water.
13. Drain off excess water, blot dry with paper towel, and let the slide air dry.
14. Find the specimen under brightfield microscope with 10X objective lens.

Fig. 6.2 ***Heat steaming use flame of burner.***

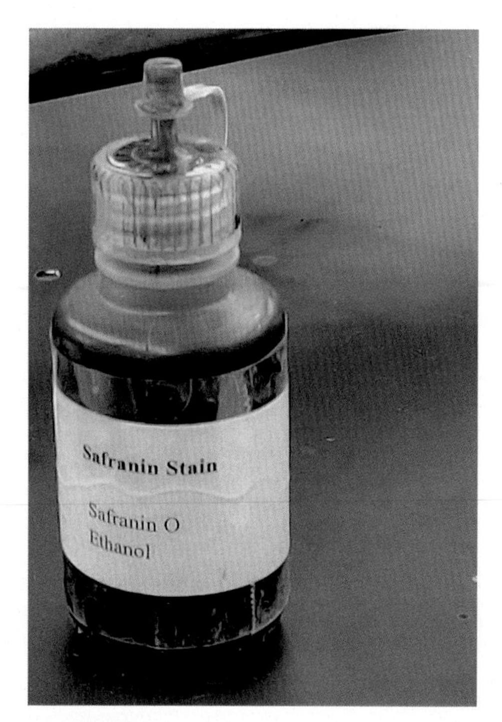

Fig. 6.3 ***Safranin reagent bottle.***

Fig. 6.4 ***Add Safranin onto the center of slide.***

15. Switch 100X objective lens with oil immersion.
16. Draw pictures, record observations, including morphology and location of the endospore in Table 6.1.

Table 6.1 Results of endospore stain.

Bacteria names	**Morphology and arrangement under 100X with oil immersion**
Escherichia coli	
Bacillus cereus	

6.4 After class questions

1. Is bacterial endosporulation a reproductive mechanism? Why?
2. Why the location of endospore is very important information?

CHAPTER 7

Aseptic technique-bacteria transfer and streak plating

Chapter outline

7.1 Materials

Nutrient broth, nutrient agar slant, nutrient agar plate, metal loops, burner, *Escherichia coli*, *Micrococcus luteus*

7.2 Introduction

To culture bacteria means to aseptically transfer and grow bacteria in a culture medium. Aseptic technique can be defined as transferring a wanted microorganism into a given medium without carrying into unwanted microorganisms, referred as contaminants. Bacterial culture medium include liquid, solid, and semi solid (5% agar used for testing bacterial motility). A nutrient broth is a typical liquid medium which contains water, beef extract, peptone, tryptone, salt all the basic ingredients for cultivating bacteria. Solid medium contains agar (a polysaccharide of the sugar galactose) or some other solidifying agent. Solid medium, such as agars on plates, has physical structure and allows bacteria to grow in physically informative ways and showing the morphology of colonies. Streak plating is an aseptic technique by streaking bacterial solutions or solid smears into agar plants to form single colonies, referred as pure culture, which is also referred as a serial dilution of bacteria culture onto solid medium. In this lab section, we will practice inoculate bacteria into a nutrient broth, onto a nutrient agar slant, and onto a nutrient agar surfaces from a pure culture in agar plates.

Introductory Microbiology Lab Skills and Techniques in Food Science
DOI: https://doi.org/10.1016/B978-0-12-821678-1.00019-8

7.3 Procedure

Exercise 1. Inoculate a nutrient broth from a plate culture

1. **Label the tube of nutrient broth:** Use permanent marker to label the top of the tube of nutrient broth including date, your initial, bench number, and name of bacteria as shown in Fig. 7.1.

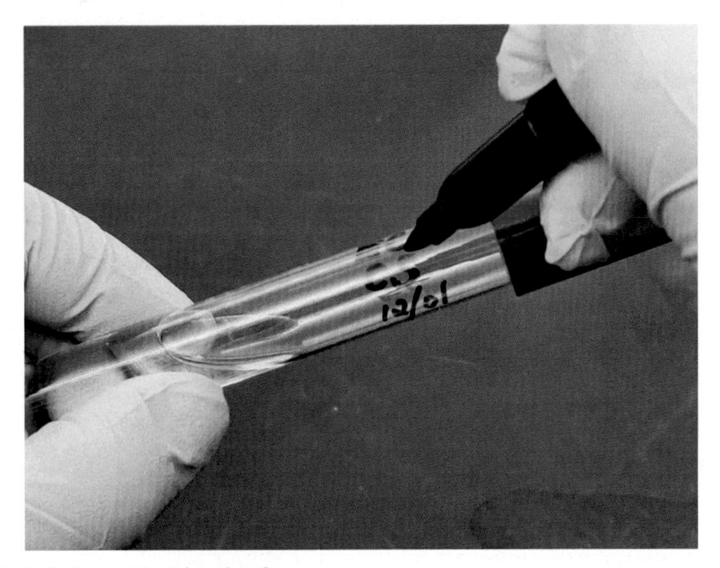

Fig. 7.1 ***Label a nutrient broth tube.***

2. **Flame loop:** Hold the bacterial metal loop like holding a pencil and heat the wire through the fires from the burner until it is red as shown in Fig. 7.2.
3. **Open the lid of a fresh nutrient broth:** Using the little finger of the same hand holding the sterilized loop to remove the lid.
4. **Flame the lip of the tube of the nutrient broth:** Gently passing the lip of the tubes through the burner as shown in Fig. 7.3.
5. **Pick a bacterial colony from the agar plate:** Place the sterilized loop onto one colony from the agar plate and remove a small amount of the culture as shown in Fig. 7.4.
6. **Transfer the bacterial colony into the nutrient broth:** Place the loop touched with the colony into the fresh nutrient broth by holding the tube horizontally as shown in Fig. 7.5 and shake slightly.
7. **Flame the lip of the tube of the nutrient broth:** Gently passing the lip of the tubes through the burner before closing the cap of the tube by little finger.

Fig. 7.2 ***Flame the loop.***

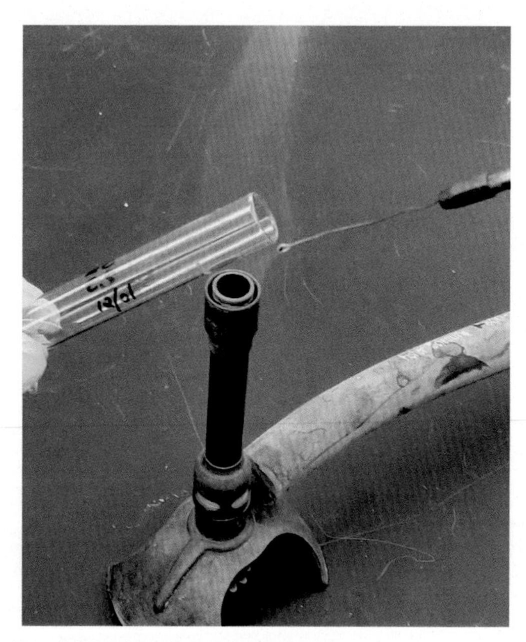

Fig. 7.3 ***Pass the lip of the tubes through the burner.***

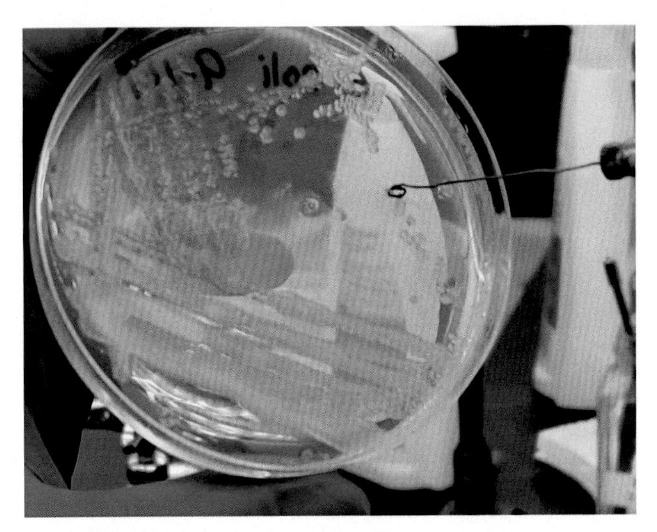

Fig. 7.4 ***Pick up bacterial colony from an agar plate.***

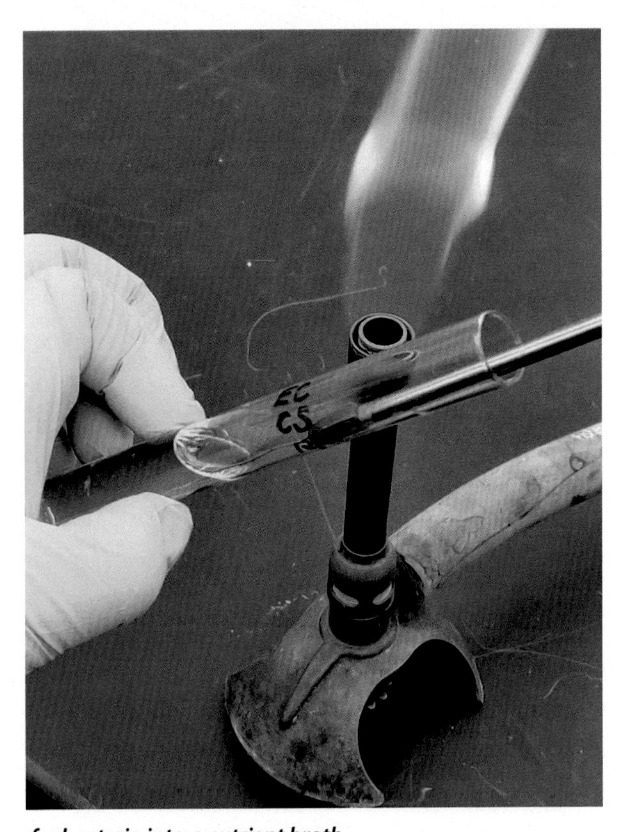

Fig. 7.5 ***Transfer bacteria into a nutrient broth.***

8. **Close the lid of the agar plates.**
9. **Flame loop again.**
10. **Incubate the slant at incubator for 35°C for 24 to 48 h.**

Exercise 2. Inoculate a nutrient agar slant from a plate culture

1. **Label the tube of nutrient agar slant:** Use permanent marker to label the side of the tube of nutrient agar slant including date, your initial, bench number, and name of bacteria as shown in Fig. 7.6.

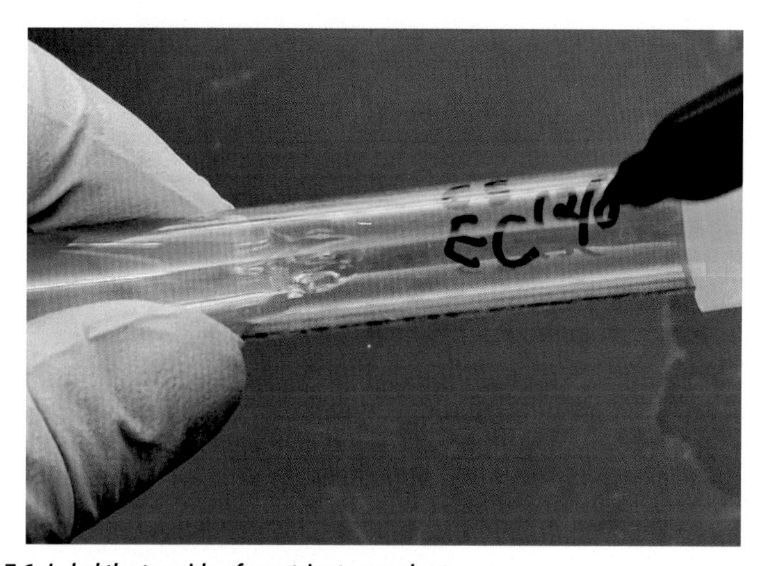

Fig. 7.6 ***Label the top side of a nutrient agar slant.***

2. **Flame loop:** Hold the bacterial metal loop like holding a pencil and heat the wire through the fires from the burner until it is red.
3. **Open the lid of a fresh nutrient agar slant:** Using the little finger of the same hand holding the sterilized loop to remove the lid.
4. **Flame the lip of the tube of the nutrient agar slant:** Gently passing the lip of the tubes through the burner.
5. **Pick a bacterial colony from the agar plate:** Place the sterilized loop onto one colony from the agar plate and remove a small amount of the culture.
6. **Transfer the bacterial colony onto the nutrient agar slant:** Place the loop touched with the colony onto the nutrient agar slant by streaking across the nutrient agar slant in a zig-zag motion to disperse the inoculum over the entire surfaces with touching the bottom liquid area as shown Fig. 7.7.

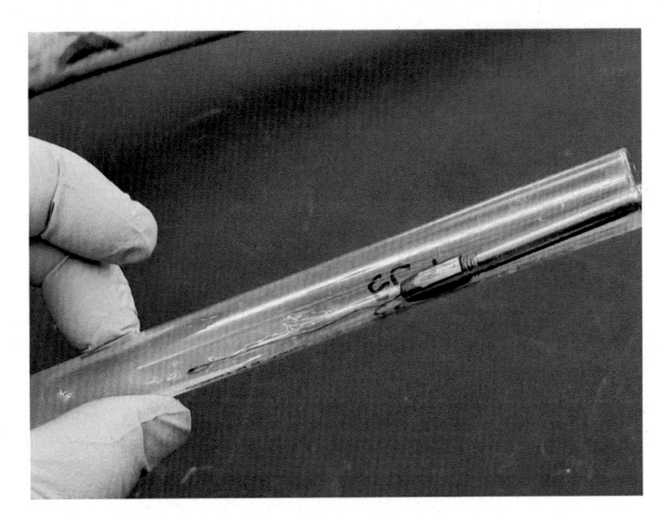

Fig. 7.7 ***Streak bacteria across the nutrient agar slant.***

7. **Flame the lip of the tube of the nutrient agar slant:** Gently passing the lip of the tubes through the burner before closing the cap of the tube by little finger.
8. **Close the lid of the agar plates.**
9. **Flame loop again.**
10. **Incubate the slant at incubator for 35°C for 24 to 48h.**

Exercise 3. Inoculate a nutrient agar plate from a plate culture by streaking-plating

1. **Label the nutrient agar plate:** Use permanent marker to label the bottom of the nutrient agar plate including date, your initial, bench number, and name of bacteria. Never label the lid!
2. **Flame loop:** Use flamed loop pick bacteria from a plate agar begin with inoculating from the first quadrant of the agar surface. Light touch without damaging the agar surfaces.
3. **Pick colony:** Flame loop, cool down by touching the uninoculated area of the agar surface.
4. **Rotating:** Rotate the plate, picking up area from the quadrant one and streak again.
5. **Streak-plating:** Flame loop, rotate plate, and repeat procedure for quadrants three and four as shown in Fig. 7.8.
6. **Incubation:** Incubate plate inverted at incubator for 35°C for 24 to 48 h.

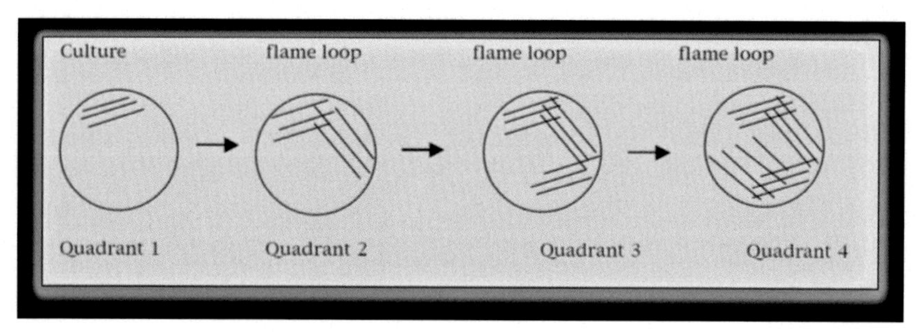

Fig. 7.8 ***Streak plating onto a nutrient agar plate.***

7.4 After class questions

1. What is a pure culture, a bacterial colony?
2. Why is it necessary to isolate individual colonies from a mixture in the medical lab?
3. Please make your own flow chart of streak plating.

CHAPTER 8

Lab practice test 1 – identify unknown bacteria using Gram-stain

Chapter outline

8.1 Introduction

This is a lab assignment for detecting unknown bacterial cultures by streak-plating and Gram staining. Students are required to complete lab report after lab practice sections. The whole assignment worth 15 points contributed to the final general microbial class total grade points.

8.2 Procedure

1. Each student obtained an unknown mixed bacterial culture and a nutrient agar plate from the front desk.
2. Label the bottom of the nutrient agar with name, bench number, and date.
3. Conduct streak plating of the unknown mixed solutions onto the nutrient agar.
4. Incubate the plates at 24°C for 48 h.
5. Observe, describe, and record the bacterial colony morphology onto the notebook.
6. Obtaining two control bacterial culture of *Micrococcus luteus* and *Escherichia coli* from the front desk.
7. Pick colony-one of the two different culture colonies from your nutrient agar plates, conduct Gram staining with the controls *M. luteus* and *E. coli* on the same slide.
8. Pick culture-two of the two different culture colonies from your nutrient agar plates, conduct Gram staining with the controls *M. luteus* and *E. coli* on the same slide.

Introductory Microbiology Lab Skills and Techniques in Food Science
DOI: https://doi.org/10.1016/B978-0-12-821678-1.00020-4

9. After compared with the Gram results of the two control cultures, figure out bacterial strains of the unknown colonies-one and -two.
10. Turn in this sheet, a brief lab report, the streak plate and the blotted slides for evaluation.
11. Complete the lab report based on the requirement below.
12. Include Table 8.1 as a front page of your lab report.

Table 8.1 Results of the unknown bacterial cultures.

Observation	Culture A	Culture B
Colony characteristics Cell shape Gram reaction		

8.3 Lab report (1–2 pages, typed)

The lab report must include the following labeled sections:

1. Title
2. Introduction: What is the purpose of this lab?
3. Materials and Methods: Describe the materials that you used and briefly outline the procedure(s).
4. Results: Record a summary of your results on the Table shown above. Provide a title for the table. Refer to this table when you write your discussion/conclusions.
5. Discussion/Conclusions: Discuss this experiment. Make sure you discuss how your results support your conclusions as Table 8.1

CHAPTER 9

Introduction of bacteria medium, nutritional requirements (synthetic and complex media), selective & differential media (MacConkey, mannitol salt, blood agar)

Chapter outline

9.1 Introduction

A bacterial culture medium is a liquid or solid preparation to cultivate, grow, transfer, and store bacteria. Based on the chemical and physical types, the medium can be categorized as synthetic (defined) medium and complex medium. Synthetic medium is a medium in which all chemical ingredients are well described as M9 medium for *Escherichia coli* and BG-11 medium for cyanobacteria. It needs to emphasize that any medium containing tryptone, peptone, and beef extract are not synthetic medium since the chemical components are not known in them. Medium that contains any ingredients with unknown chemical components are called complex medium. Based on the functional types, the medium can be categorized as supportive medium, selective medium, and differentiate medium. Supportive medium, such as nutrient broth, tryptic soy broth, and tryptic soy agars can cultivate many bacteria in the exception of fastidious bacteria. Selective medium containing certain ingredients (bile salts, crystal violets) favor the growth of a particular bacteria or a certain group of bacteria.

Introductory Microbiology Lab Skills and Techniques in Food Science
DOI: https://doi.org/10.1016/B978-0-12-821678-1.00010-1

Differential medium are the media that distinguish different groups of bacteria based on their biological characteristics typically biochemical reaction resulting color change of colonies. Below we will introduce three selective and/or differentiate medium. *Note:* A bacterial culture medium can be fit into different categories such as MacConkey agar is a both selective and differential medium.

9.2 MacConkey agar

It contains lactose, neutral red, crystal violet, and bile salts. The crystal violet and bile slats inhibit the growth of gram-positive organisms but support the growth of gram-negative organisms. The presence of lactose and neutral red (a pH indicator) allows the differentiation of gram-negative bacteria whether they use lactose during fermentation. If the acid products released during the lactose fermentation, the neutral red turns red color. For example, colonies of *E. coli* O157:H7 are pink/red (due to lactose fermentation) surrounded by a bile salts precipitation zone. Non-O157 shiga-toxin-producing *E. coli* colony is colorless due to the nonlactose fermentation.

9.3 Mannitol salt agar

It contains 7.5% salt, 6-carbon mannitol sugar, and the pH indicator phenol red. The 7.5% salt inhibit most bacteria other than staphylococci. Mannitol can be fermented to generate the yellow colonies due to the phenol red turns yellow color for the generation of acidic products. For example, colonies of *Staphylococcus aureus* are yellow due to the mannitol fermentation. *Staphylococcus epidermis* colony is colorless due to the nonmannitol fermentation.

9.4 Blood agar

It is a tryptic soy agar containing 5% sheep blood can support most bacteria grow dramatically and differentiate growth of bacteria based on the hemolysis of the red blood cells caused by hemolysin. A clear zone around the colony indicates a complete hemolysis referred as β-hemolytic (*Streptococcus pyogenes*), or a greenish halo around the colony suggests an incomplete hemolysis referred as α-hemolytic (Streptococcus *pneumonia*), if no change of the colony suggests a non-hemolysis referred as γ-hemolytic (*Enterococcus faecalis*).

9.5 Procedure

1. **Label the bottom of the agars:** Split the bottom of the MacConkey, mannitol salt, and blood agars with three quarters using permanent marker with the initial name of the bacteria on each quarter as "EC" referring *E. coli*, "PA" referring *Pseudomonas aeruginosa*, and "SA" referring *S. aureus*. Label date, your name initial, and bench number of the side of the bottom as shown in Figs. 9.1 and 9.2.

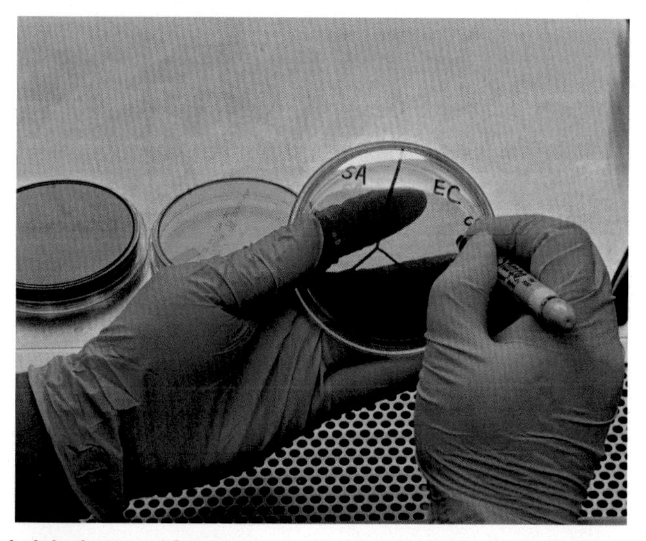

Fig. 9.1 ***Label the bottom of agars.***

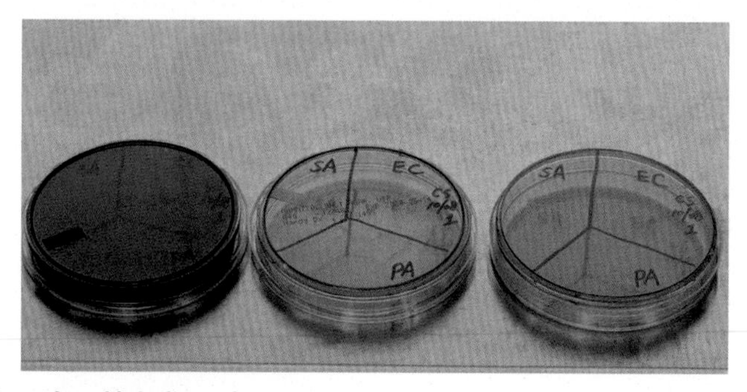

Fig. 9.2 ***Completed labeling of agars.***

2. **Spot inoculation:** Use your loop choose bacterial colonies of the assigned bacteria from the agar onto each section with gentle streak as shown in Figs. 9.3 and 9.4. Flame loop between picking up different bacteria to prevent cross-contamination.

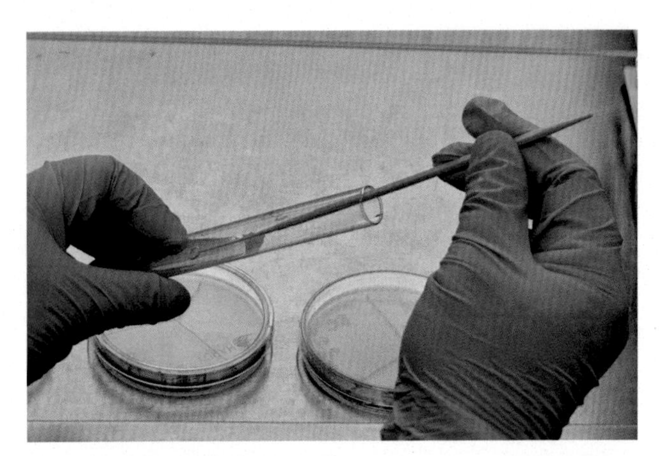

Fig. 9.3 ***Pick bacteria from agar slant.***

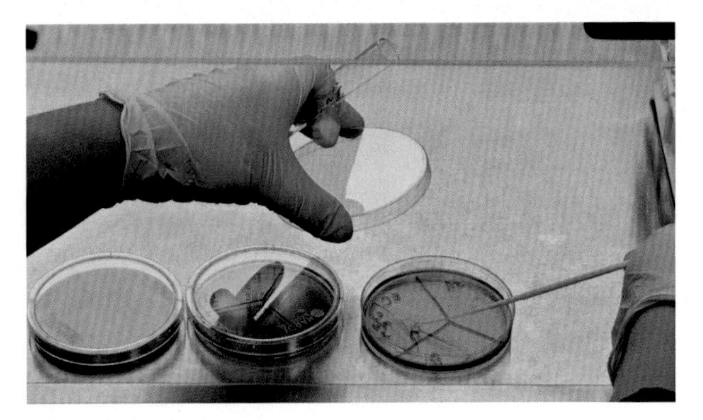

Fig. 9.4 ***Spot inoculation into agars.***

3. **Incubation:** Incubate agar plates inverted at 35°C for 24–48 h.
4. **Record** the bacterial growth situation in Table 9.1.

Table 9.1 Results of bacteria growth on selective and differential medium.

	Escherichia coli	***Pseudomonas aeruginosa***	***Staphylococcus aureus***
MacConkey Agar			
Mannitol salt Agar			
Blood agar			

9.6 After class questions

1. Why blood agar is a differential medium?
2. Explain the selective nature of MacConkey and mannitol salt agars.

CHAPTER 10

Lab practice test 2 – dilution technology and quiz

Chapter outline

10.1 Introduction

Bacterial vegetative cells growing in broth liquid solution, such as nutrient solution and tryptic soy broth, can reach the population of approximately 10^8–10^9 cells/mL (8.0–9.0 $\log_{10}$ CFU/mL) after incubating at 35°C for 24–48 h. This amount of bacterial population is too many to be on agar medium to count and calculate the bacterial concentrations (CFU/mL) in the liquid solution. Therefore, in order to calculate the concentration of bacterial cells, the 10-fold or 100-fold serial dilution were needed to be done in buffered bacterial solutions (buffered peptone water) before adding 0.1 or 1.0 ml of the solution onto the agar medium by spreading or pour plating. The prepared sterilized 9.0 mL or 9.9 mL 0.1% buffered peptone water were used for 10-fold and 100-fold serial dilution.

10.2 Dilution technique figures (assume 0.1 mL of each dilution tube adding onto agar plates)

1. 10-fold serial dilution as shown in Fig. 10.1.
2. 100-fold serial dilution as shown in Fig. 10.2.
3. A combination of 10-fold and 100-fold serial dilution as shown in Fig. 10.3.

Note: "0" dilution-averagely spreading 1.0 mL original solution onto three agar plates, adding all colony-forming unit (CFU) of three agars after incubation.

Introductory Microbiology Lab Skills and Techniques in Food Science
DOI: https://doi.org/10.1016/B978-0-12-821678-1.00001-0

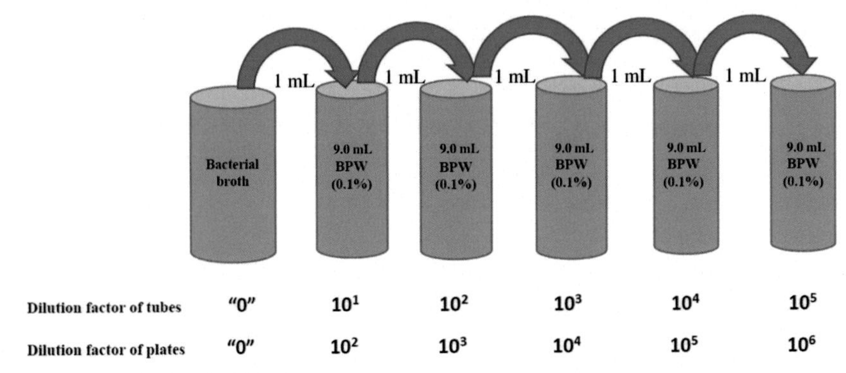

Fig. 10.1 ***10-fold serial dilution.***

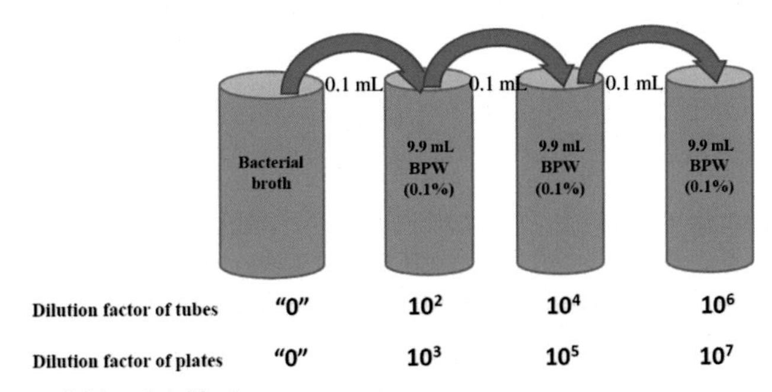

Fig. 10.2 ***100-fold serial dilution.***

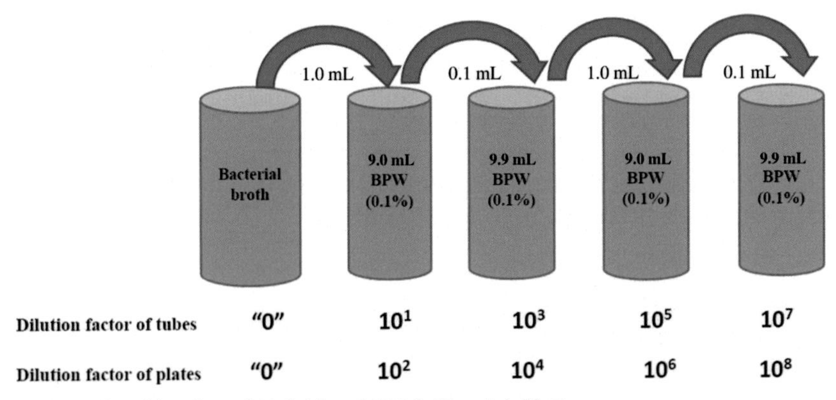

Fig. 10.3 ***Combination of 10-fold and 100-fold serial dilution.***

10.3 Procedure of calculating bacterial concentration

1. Manually count the agar plates after incubation.
2. Pick the agar plates with the counts of CFU between 30–300 CFU/plate, referred as acceptable and countable zone. 2.1. If colonies < 30, count and record it, but it will not be used for calculating the bacterial concentration, because colonies may come from contamination. 2.2. If colonies are >300, record as TNTC, referred as too-numerous-to-count.
3. The final counts are the result of final colony number of the plate × final dilution factor.
4. For example, 50 colonies on the plate with final dilution factor 10^6, then the final counts are $50 \times 10^6 = 50{,}000{,}000$ CFU/mL transferred as 7.69 $\log_{10}$CFU/mL.

10.4 After class questions

A reminder: One part of culture placed in nine parts of diluent is called a 1/10 or 10^{-1} dilution. If you take one part of the 1/10 dilution and place it in nine parts of diluent, this is an additional 1/10 dilution. This is because the final (total) dilution is a multiple of all the dilutions. In this example, the final dilution is $1/10 \times 1/10 = 1/100$ (or 10^{-2}). The denominator of the final dilution (100 or 10^2) is called the dilution factor. To determine the final dilution, multiply all the dilution factors.

1. You want to prepare 10 mL of a 1/100 dilution. What should the aliquot and diluent volumes be?
2. What is the dilution factor when you add 2 mL sample to 8 mL diluent?
3. You diluted a bacterial culture a million-fold. You then plated out 0.1 mL of the dilution and after incubation, 45 colonies appeared on the plate. How many bacteria/mL were in the original undiluted culture?
4. A sample was diluted by placing a 0.1 mL aliquot into 0.9 mL of diluent, and 1 mL of this dilution was pour plated. After incubation, 150 colonies appeared. What is the CFU/mL in the original culture?
5. A pure bacterial culture was diluted by adding a 0.1 mL aliquot to 0.9 mL water. Then, 0.1 mL of this dilution was plated out, yielding 82 colonies. Calculate the CFU/mL in the original culture.

CHAPTER 11

Numeration bacteria population by pour plating

Chapter outline

11.1 Materials

Petri-dish, melted agar, 15 mL dilution tubes, 9.9 mL and 9.0 mL 0.1% buffered peptone water, *E. coli* culture solution, 0.1 mL pipette tips, 1 mL pipettes

11.2 Introduction

Pour plating is conducted by pouring 20–25 mL of melted agar onto an empty petri dish followed by rotating the petri dish to thoroughly mixing agar medium. Heat sensitive bacteria will be killed by the mild heat (~50°C) from the melted agars, therefore, the number of colonies of pour plating will be slightly lower than spread plating. After pour plating, the agars will be stored at room temperature to solidify and then incubated at certain temperature after certain time period followed by manually counting the bacterial colony-forming unit (CFU) throughout the agars and calculate the bacterial population in original solutions as CFU/mL. Since melted agars are used for pour plating, which gives us an opportunity to practice making bacterial agar medium followed by autoclave sterilization.

11.3 Dilution schedule as shown in Fig. 11.1

Procedure:

1. **Preparing melted bacterial agar medium:** Add 30 gm of tryptic soy agar power into 1L distilled water, mixed very well with electric magnetic stir, followed by covering the cup of glass bottles with foil

Introductory Microbiology Lab Skills and Techniques in Food Science
DOI: https://doi.org/10.1016/B978-0-12-821678-1.00018-6

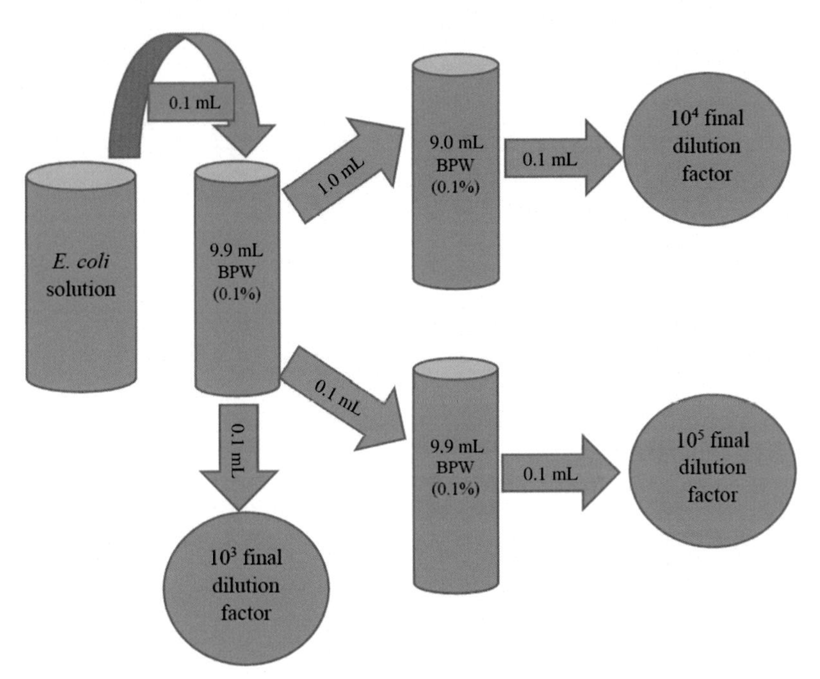

Fig. 11.1 ***Dilution schedule of pour plating.***

paper with attached autoclave taps. The bottles will be autoclaved under liquid cycle at 121°C for 15 min followed by put autoclaved bottles into water bath with temperature set at 50°C.

2. **Label petri dishes:** Collect new empty petri-dish plates and clearly label at the bottom of the agar as initial, bacterial name, final dilution factors as "10^3," "10^4," and "10^5," and the date as shown in Fig. 11.2.

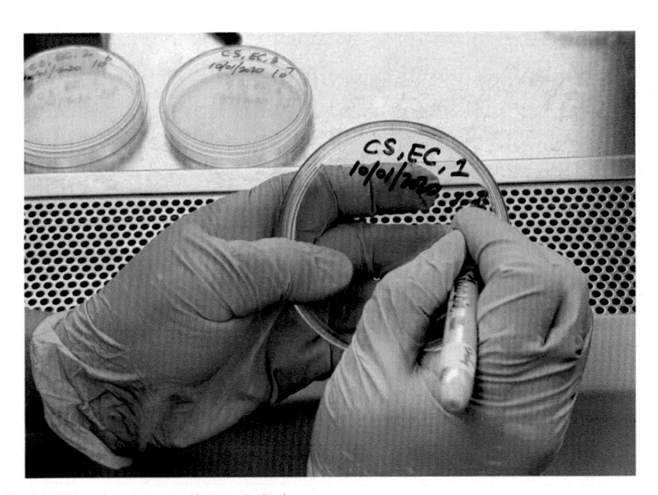

Fig. 11.2 ***Label the bottom of petri dish.***

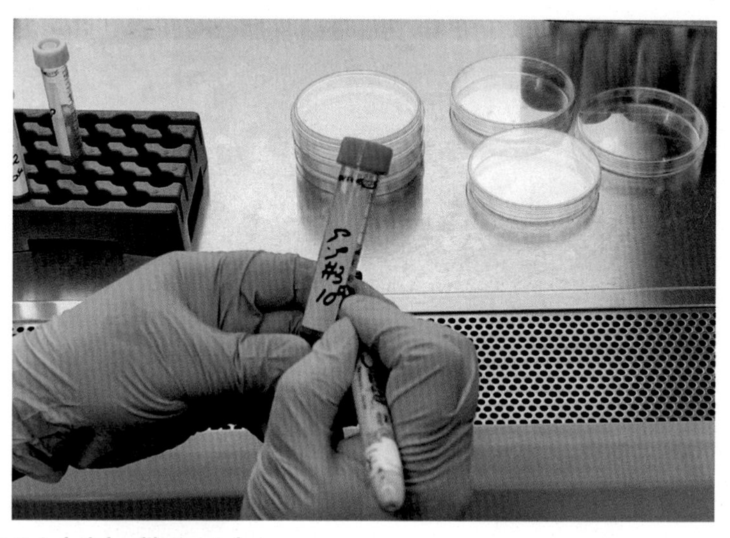

Fig. 11.3 ***Label the dilution tube.***

3. **Label dilution tubes:** Collect 2 9.9 mL 0.1% buffered peptone water (BPW) solution and 1 9.0 mL BPW solution tubes. Label the tubes of 9.9 mL as "#1 10^1" and "#2 10^4." Label the tube of 9.0 mL as "#1 10^3" as shown in Fig. 11.3.
4. **Make the 100-fold or 10-fold serial dilution:** Use 0.1 mL or 1.0 mL of pipettes to transfer the bacterial solution (24h fresh *E. coli* culture solution) followed by the dilution schedule as shown in Fig. 11.1. Change new pipettes between tubes to prevent cross contamination, and mixing each dilution tube very well as shown in Fig. 11.4.
5. **Add diluted solutions onto petri dishes:** Organize the petri-dish plates from the greatest to the lowest final dilution factor. Use 0.1 mL of pipettes to add 0.1 mL of diluted solutions onto the center of the new empty petri-dish as shown in Fig. 11.5. Always add as the order from the lowest to greatest dilution plates, which only need one pipette.
6. **Pour plating:** Add approximately 20–25 mL of prepared melted agar solution onto the petri dish. Gently rotating the agar plates for 1 min as shown in Fig. 11.6.
7. **Solidify agars:** Let the agars stay at room temperature for 30 min to 1 h to completely solidify the agars.
8. **Incubation:** Stack the agar plates upside-down and incubate at 35°C for 24 to 48 h as shown in Fig. 11.7.

Fig. 11.4 *Mix the dilution tube.*

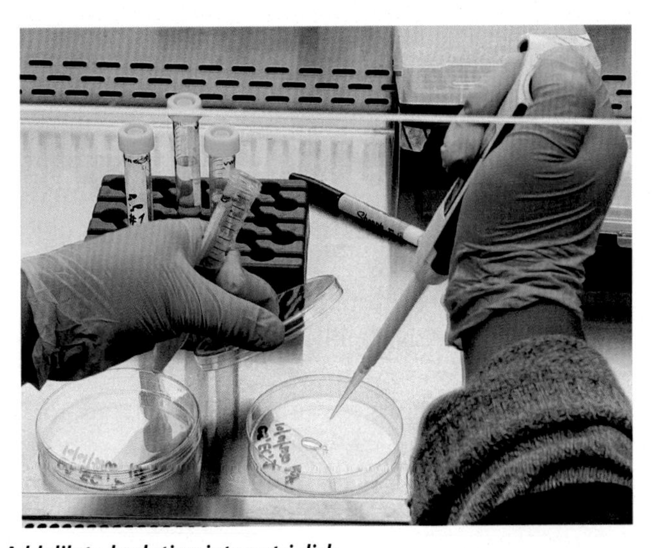

Fig. 11.5 *Add diluted solution into petri dish.*

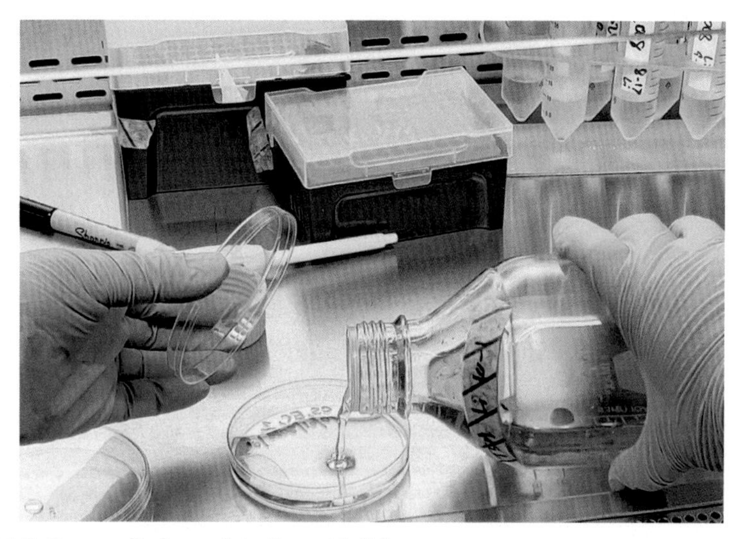

Fig. 11.6 ***Pour melted agar into the petri dish.***

Fig. 11.7 ***Put agars into the incubator.***

9. **Count and calculate of bacterial solutions:** Organize the plates from the lowest to greatest dilution factor, choose the plates with estimated CFU counts between 30–300, then manually count the colonies by using markers to mark the colonies already been counted. Calculate the bacterial concentrations as CFU of the plates × final dilution factor of the agar plates.

11.4 After class question

What is the advantage of pour plating to numeration and calculation bacterial population?

CHAPTER 12

Numeration bacteria population by spread plating

Chapter outline

12.1 Materials

Nutrient agar, spreader, 15 mL dilution tubes, 9.9 and 9.0 mL 0.1% buffered peptone water, Pipette tips, Pipette, *Escherichia coli* solution

12.2 Introduction

Spreading plating is conducted by spread liquid solution on agar medium using sterilized plastic spreader or flamed glass spreader. After spread plating, the agars will be stored convertibly (upside down) and incubated at certain temperature after certain time period followed by manually counting the bacterial colony-forming unit (CFU) on agars and calculate the bacterial population in original solutions as CFU/mL.

12.3 Dilution schedule as shown in Fig. 12.1

Procedures

1. **Label agar plates:** Collect three agar plates and clearly label at the bottom of the agar as initial, bacterial name, final dilution factors as "10^5," "10^6," and "10^7," and the date as shown in Fig. 12.2.
2. **Label dilution tubes:** Collect 3 9.9 mL 0.1% buffered peptone water (BPW) solution and 1 9.0 ml BPW solution tubes. Label the tubes of 9.9 mL as "#1 10^2," "#2 10^4," and "#3 10^6." Label the tube of 9.0 ml as "#1 10^5".

Introductory Microbiology Lab Skills and Techniques in Food Science
DOI: https://doi.org/10.1016/B978-0-12-821678-1.00007-1

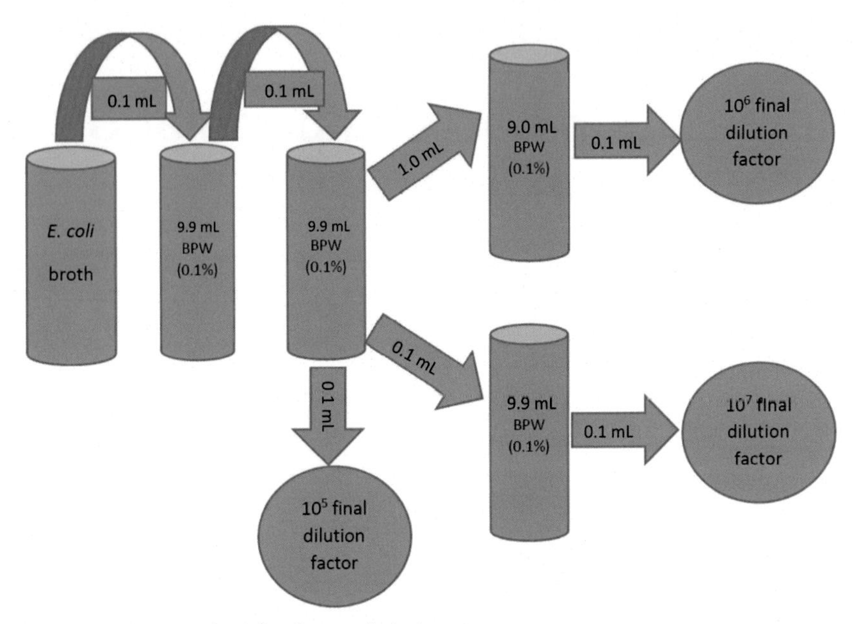

Fig. 12.1 ***Dilution schedule of spread plating.***

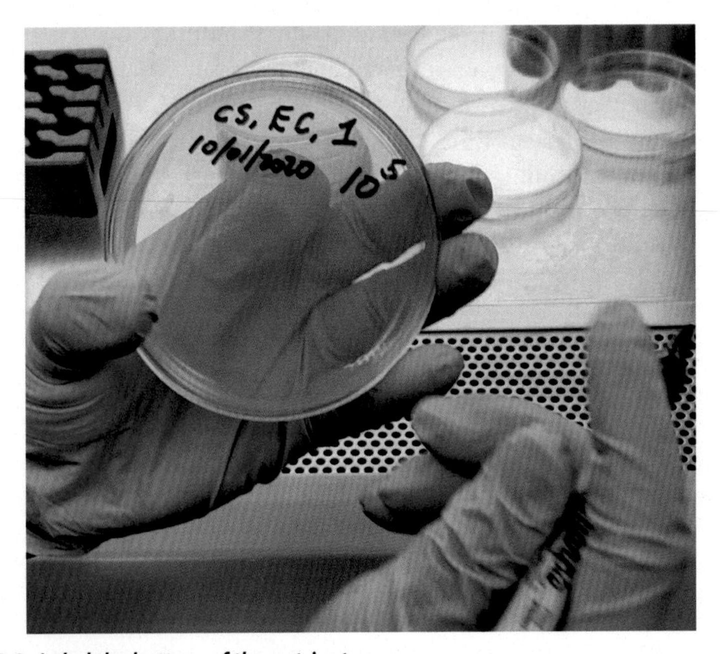

Fig. 12.2 ***Label the bottom of the nutrient agar.***

3. **Make the 100-fold or 10-fold serial dilution:** Use 0.1 mL or 1.0 mL of pipettes to transfer the bacterial solution (24 h fresh *E. coli* culture solution) followed by the dilution schedule as shown in Fig. 12.1. Change new pipettes between tubes to prevent cross-contamination. Remember, the dilution factor of tubes is always 10 times greater than the final dilution factor of the agar plates.
4. **Add diluted solutions onto agar plates:** Organize the agar plates from the greatest to the lowest final dilution factor. Use 0.1 mL of pipettes to add 0.1 mL of diluted solutions onto agar surfaces as shown in Fig. 12.3. Always add as the order from the lowest to greatest dilution plates, which only need one pipette.

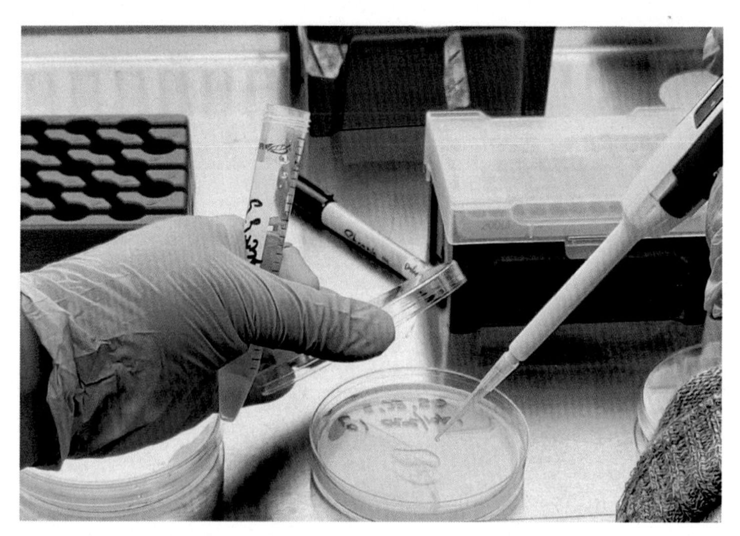

Fig. 12.3 ***Add 0.1 mL diluted solution onto the nutrient agar.***

5. **Spread plating:** Use plastic sterilized spreader to averagely spread solutions on the agar surfaces as shown in Fig. 12.4. Let them dry out for 3–5 min.
6. **Incubation:** Stack the agar plates upside down and incubate at 35°C for 24 to 48 h.
7. **Count and calculate of bacterial solutions:** Organize the plates from the lowest to greatest dilution factor, choose the plates with estimated CFU counts between 30 and 300, then manually count the colonies by using markers to mark the colonies already been counted. Calculate the bacterial concentrations as CFU of the plates × final dilution factor of the agar plates.

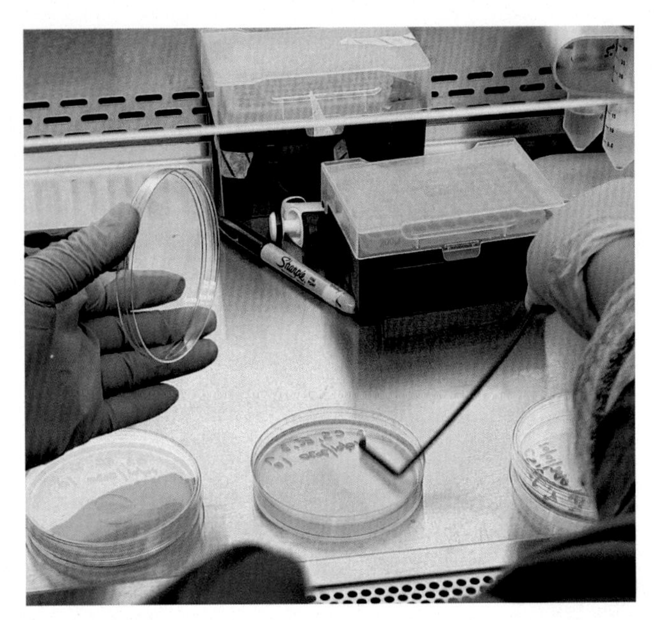

Fig. 12.4 ***Spread plating.***

12.4 After class question

What is the advantage of spread plating to numeration and calculation bacterial population?

CHAPTER 13

External conditions for bacterial growth (temperature, pH, oxygen, osmotic)

Chapter outline

13.1 Materials

Nutrient agar slants, thioglycolate broth, *Geobacillus, Staphylococcus aureus, Escherichia coli, Serratia marcescens*

13.2 Introduction

Growth and survival of bacterial cells are affected by the environmental factors surrounding them. Based on the requirement of optimal growth temperatures, bacteria can be categorized as psychrophile, psychrotroph, mesophile, thermophile, and hyperthermophile. Based on the optimal oxygen level, bacteria can be categorized as aerobic, facultative, and anaerobic. Based on the osmatic conditions, bacteria can be categorized as nonhalophiles, halophiles, and extreme halophiles. The Table 13.1 summarize the impact of these environmental factors on bacterial growth.

13.3 Exercise 1. Optimal growth requirements (temperature)

Organisms to be investigated

1. ***Serratia marcescens*:** This organism has the interesting property of displaying a different phenotype depending upon the temperature at which it is grown.

Introductory Microbiology Lab Skills and Techniques in Food Science
DOI: https://doi.org/10.1016/B978-0-12-821678-1.00009-5

Table 13.1 Summary of environmental factors on bacterial growth.

Environmental factors	Categorize	Growth requirements
Temperature	Groups	Growth temperature ranges
	Psychrophile	0°C to 20°C
	Psychrotroph	0°C to 35°C
	Mesophile	20°C to 45°C
	Thermophiles	55°C to 85°C
	Hyperthermophiles	85°C to 113°C
Oxygen	Groups	Requirement of O_2
	Aerobic	Completely needs O_2 for growth
	Facultative	No require but grow better with O_2
	Anaerobic	Dies in the presence of O_2
Osmotic	Groups	Requirement of salt concentrations
	Nonhalophiles	Grow optimally at < 0.2 M
	Halophiles	Grow optimally at 0.2–2.0 M
	Extreme halophiles	Require >2 M
	Groups	Requirement of pH
pH	Acidophile	0–5.5
	Neutrophile	5.5–8.0
	Alkaliphile	8.0–11.5

2. **Geobacillus stearothermophilus:** This organism is a thermophile and its optimum growth temperature would occur between 50–80°C.
3. ***Escherichia coli*:** This organism has an optimal growth temperature range of 37°C.
4. ***Staphylococcus aureus:*** The optimal growth temperature of *S. aureus* is 35–37°C. Please study lecture material to investigate the heat susceptibility of the major toxins produced by *S. aureus*.

Nutrient agar slants (4/group of two students):

1. **Label:** Label each slant as shown in Fig. 13.1 with your assigned organism (either *Geobacillus, S. aureus, E. coli, S. marcescens*), and your bench partner initials, the date and one of the following incubation temperatures (1) 4°C, (2) 25°C, (3) 37°C, or (4) 60°C.
2. **Inoculation:** Inoculate each slant with your assigned organism as shown in Fig. 13.2.

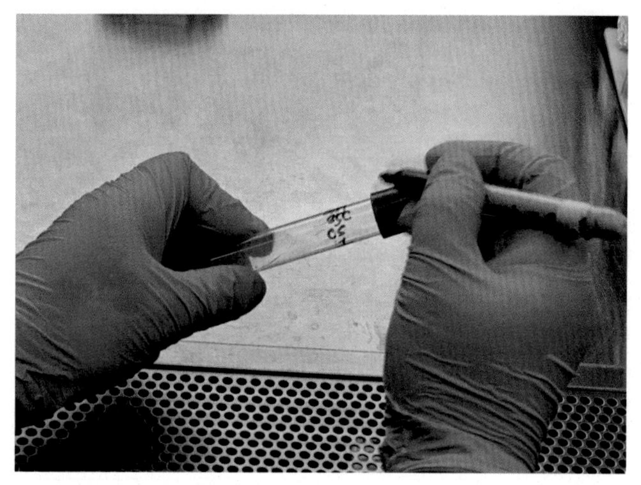

Fig. 13.1 ***Label each nutrient slant.***

3. **Incubation:** Place the inoculated slants into the appropriate racks that have been labeled with the incubation temperature; you will find these racks on the front desk.
4. **Observation:** After 24 h incubation, the tubes that appear to have no growth should be compared to a tube containing sterile nutrient broth.

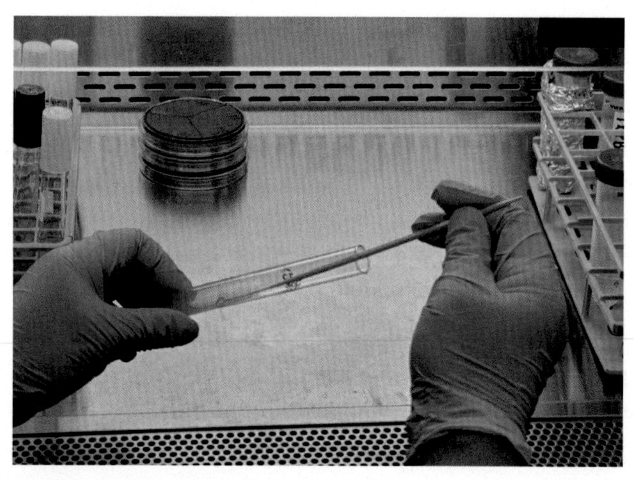

Fig. 13.2 ***Inoculate a nutrient slant.***

13.4 Exercise 2. Physiological effects of temperature

In this exercise, we will investigate an example of how temperature affects physiological reactions (pigment production).

Procedure:

1. **Label slants (2/group of two):** Label 1 slant *S. marcescens*, 25°C; Label 1 slant *S. marcescens*, 37°C.
2. **Inoculation:** Inoculate slants using a sterile loop to transfer *S. marcescens* from the broth cultures prepared for you.
3. **Incubation:** Place the inoculated tubes into the appropriate racks that have been labeled with the incubation temperature; you will find these racks on the front desk.
4. **Observation:** After 24 h incubation, read and compare the nutrient agar slants of *S. marcescens*. Record your observations in your lab notebook.

13.5 Exercise 3. Cultivation of aerobic, facultative, and anaerobic bacteria in thioglycolate broth

Thioglycolate broth is a nutrition broth containing a reducing agent sodium thioglycolate to removing oxygen from the broth. Methylene blue as a chemical indicator is included in the broth to indicate the presence of oxygen when it shows blue color.

Procedure:

1. **Label tubes (3/per student):** Obtain three thioglycolate broth tubes (do not shake these tubes!) and label each tube for assigned bacterial cultures as *Clostridium perfringens*, *S. aureus*, and *Pseudomonas aeruginosa*.
2. **Inoculation:** Inoculate the three broth tubes using a sterile loop to transfer the bacteria from the slants prepared for you.
3. **Incubation:** Incubate the thioglycolate broth at 37°C for 24 to 48 h.
4. **Observation:** After 24–48 h incubation read and compare the three thioglycolate broth tubes. Record your observations in your lab notebook.

13.6 Exercise 4. Impact of osmotic pressure on bacterial growth

Procedures

1. **Label tubes (5/per student):** Obtain five nutrient agar slants each containing 0.85%, 1%, 2%, 5%, and 10% of salt (sodium chloride) and label the salt concentrations.

2. **Inoculation:** Inoculate your assigned culture *E. coli*, *S. aureus*, or *Halobacterium salinarium* the five slants using a sterile loop to transfer the bacteria from the agar plates prepared for you.
3. **Incubation:** Incubate the slants at 37°C for 24 to 48 h.
4. **Observation:** After 24–48 h incubation read and compare the five slants. Record your observations in your lab notebook.

13.7 Exercise 5. Impact of pH on bacterial growth

Procedure:

1. **Label tubes (5/per student):** Obtain five nutrient broth each with pH values of 3.0, 5.5, 7.0, 8.5, 10.0 and label the pH values.
2. **Inoculation:** Inoculate your assigned culture *E. coli*, *S. aureus*, or *Bacillus cereus* the five nutrient broth using a sterile loop to transfer the bacteria from the agar plates prepared for you.
3. **Incubation:** Incubate the broth at 37°C for 24 to 48 h.
4. **Observation:** After 24–48 h incubation read and compare the five broth. Record your observations in your lab notebook.

Note: To prepare nutrient broth with pH values of 3.0, 5.5, 7.0, 8.5, and 10.0, it needs to be pretested the amount of HCl and NaOH adding into the broth solutions after autoclave to achieve the targeted pH values. The solutions cannot be preadjusted for pH values followed by autoclave because autoclave with high pressure and high temperature could dramatically shift the targeted pH values.

13.8 After class questions

1. Define psychrophile, mesophile, thermophile, hyperthermophile (refer to your lecture slides).
2. Why is it important to know the optimum growth temperature of microbes?
3. Describe the most common anaerobic pathogen and the types of disease they cause?
4. Define nonhalophiles, halophiles, and extreme halophiles (refer to your lecture slides).
5. Define acidophile, neutrophile, alkaliphile (refer to your lecture slides).
6. Why *S. marcescens* at 25°C showing pink color as shown in Fig. 13.3?

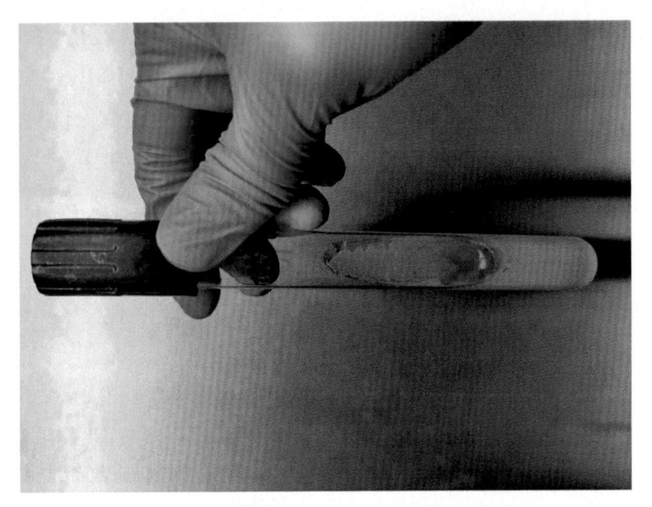

Fig. 13.3 ***Serratia marcescens* at 25°C.**

CHAPTER 14

Cultivation of anaerobic bacteria

Chapter outline

14.1 Introduction

Based on the requirement of oxygen demand, bacteria can be categorized as anerobic, anaerobic, and facultative. Aerobes are bacteria requiring oxygen at their terminal electron acceptor. Anaerobes cannot survive in the presence of oxygen due to the lack of enzymes like superoxide dismutase (SOD) and catalase and they use other inorganic chemicals such as nitrate, nitrite, and sulfites as their terminal electron acceptor. Most bacteria are facultative that they can grow with or without oxygen but grow better with oxygen. *Clostridium perfringens* is an anaerobic with endospore-forming bacteria and typically contaminated with canned foods referred as low acid food (pH ~4.6, and water activity ~0.86). Trouble signs of the canned foods indicating bacterial contamination include clear liquids turn milky, cracked jars, loosen or dented jars, swollen or dented cans and "off" odor.

Anaerobic jar as shown in Fig. 14.1 with gas generator GasPak is the basic anaerobic techniques used to create anaerobic environment in the microbial lab settings. They reduce the oxygen amount and replace it with a CO_2 atmosphere in the jar. In an anaerobic jar, the addition of 10 mL of water into a GasPak generate the reaction between sodium borohydride and sodium bicarbonate to produce hydrogen and CO_2 gas. The lid of the anaerobic jar containing a small packet of palladium pellets to act as a catalyst to let the reaction happened more efficiently. An oxygen indicator methylene blue is also included in the jar to monitor the oxygen level with blue color showing the presence of oxygen and colorless indicting the absence of oxygen.

Introductory Microbiology Lab Skills and Techniques in Food Science
DOI: https://doi.org/10.1016/B978-0-12-821678-1.00024-1

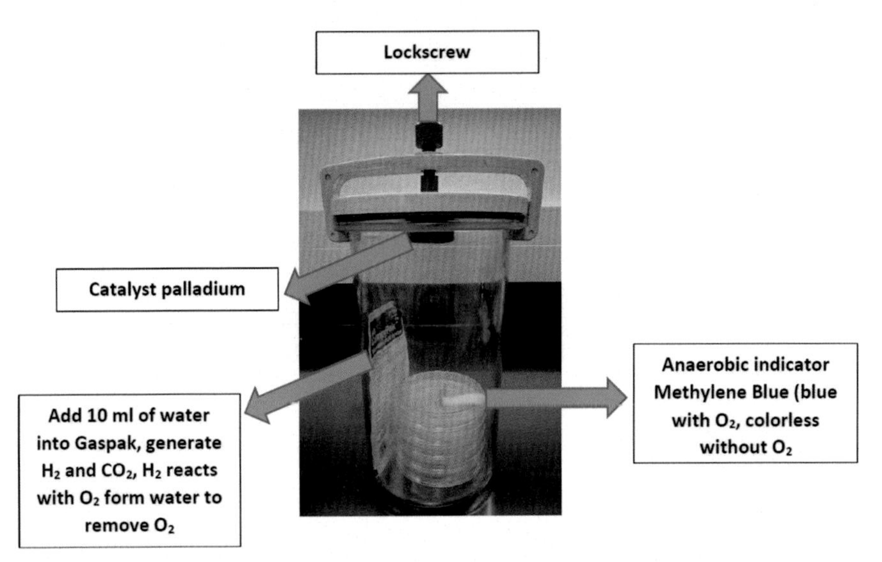

Fig. 14.1 ***Principles of anaerobic jar.***

14.2 Procedure

1. Collect the two blood agars from the front desk.
2. Label the back of the plate as "aerobic" and "anaerobic" with other basic information.
3. Using sterilized loop to streak-plate each blood agar with *Clostridium perfringens*.
4. Put one of the two blood agars into an anerobic jar, the other blood agar left in the general wood container as a control.
5. Open the GasPak by adding 10 mL of distilled water and put into the anaerobic jar.
6. Seal the lid of the anaerobic jar immediately.
7. Incubate both plates at 35°C for 48 h.
8. Observe the two plates and record the results on the notebook.
9. *Clostridium perfringens* should indicate double β-hemolytic zone on the blood agar.

14.3 After class questions

1. Describe in detail the most common anaerobic bacterial pathogen in the clinical lab and the type of disease they cause.
2. Describe the mechanism of GasPak to generate anerobic environment.

CHAPTER 15

Biochemistry test of bacteria-1 (urease test, carbohydrate fermentation, catalase test, oxidase test)

Chapter outline

The following bacteria will be assigned to each student: *Staphylococcus aureus*, *Escherichia coli*, *Bacillus subtilis*, *Proteus vulgaris*, *Pseudomonas aeruginosa*, *Enterobacter aerogenes*.

15.1 Exercise 1. Urease test: One urea broth tube/student

15.1.1 Introduction

The enzyme urease will split the urea molecule releasing carbon dioxide and ammonia. Phenol red has been added to the urea broth in order to detect the change in pH upon the release of ammonia. The test for urease does not require a reagent since the alkaline by-product ammonia will turn the broth a bright pink color if urease is produced. A clinical fact is that *Helicobacter pylori* will produce urease in the stomach to neutralize acid and allow growth which may result in the production of gastric ulcers.

Procedure:

1. Label and inoculate a urea broth with your assigned species as shown in Fig. 15.1, you will use a different species than your lab bench partner.

Introductory Microbiology Lab Skills and Techniques in Food Science
DOI: https://doi.org/10.1016/B978-0-12-821678-1.00005-8

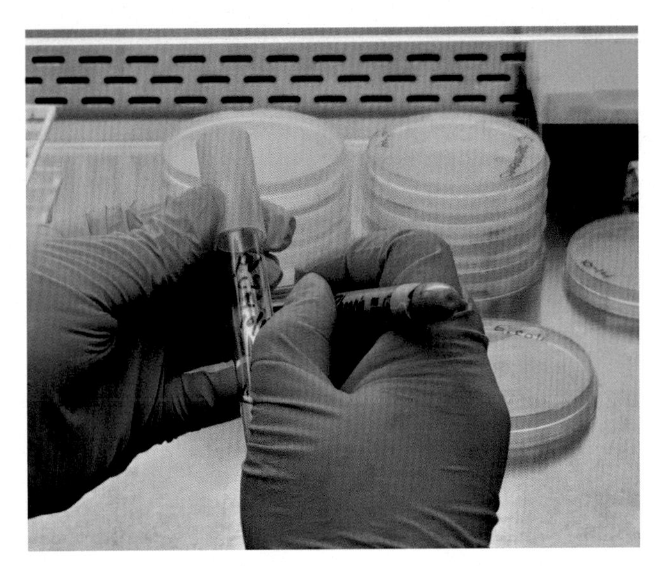

Fig. 15.1 ***Label the urea test tube.***

2. Incubate the test tube by using sterilized loop picking single colonies from the agar plate as shown in Fig. 15.2 and in Fig. 15.3.
3. Incubate at 37°C for 24 h.
4. Observe your urea broth tube of producing urease. Pink is positive.

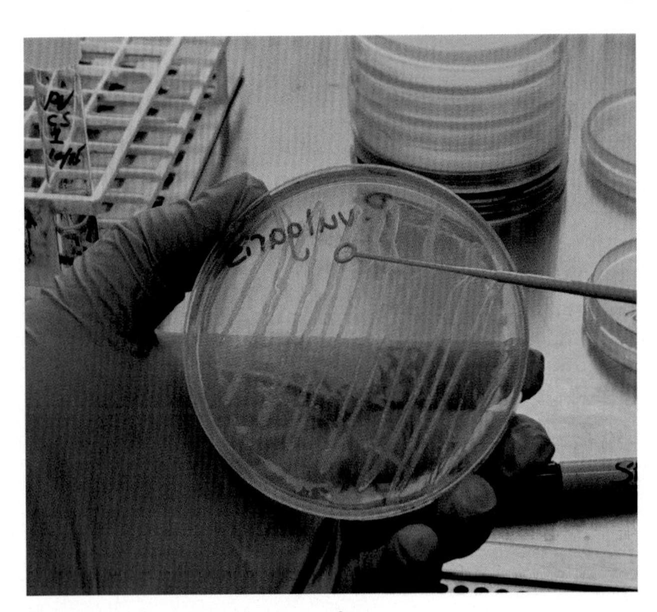

Fig. 15.2 ***Pick bacteria from nutrient agar plate.***

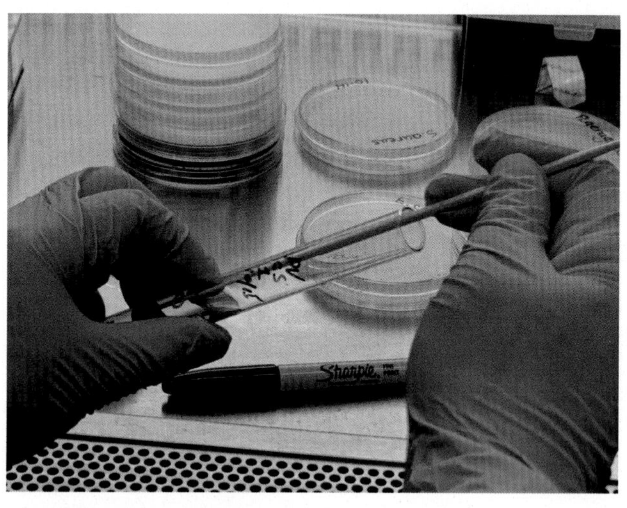

Fig. 15.3 ***Inoculate the urea test tube.***

15.2 Exercise 2. Fermentation of carbohydrates: One of each carbohydrate fermentation tube/student

15.2.1 Introduction

Media for testing carbohydrate fermentation are often prepared as tubed broths, each tube containing a small inverted "fermentation" (or Durham) tube for trapping any gas formed when the broth is inoculated and incubated. Each broth contains essential nutrients, a specific carbohydrate, and Phenol Red to indicate a change in pH if acid or base is produced in the culture (the broth is adjusted to a neutral pH when prepared). Organisms that grow in the broth but do not ferment carbohydrates produce no change in the color of the medium and no gas is formed. Some organisms may produce acid products in fermenting the sugar, but no gas, whereas other may form both acid and gas. In some cases, organisms that do not ferment the carbohydrate using the protein nutrients in the broth, thereby producing alkaline end products, a result that is also evidenced by a change in indicator color.

Procedure:

1. Label and Inoculate the following fermentation tubes with your assigned species as shown in Fig. 15.4.
   - **a.** Phenol red glucose broth
   - **b.** Phenol red sucrose broth
   - **c.** Phenol red lactose broth

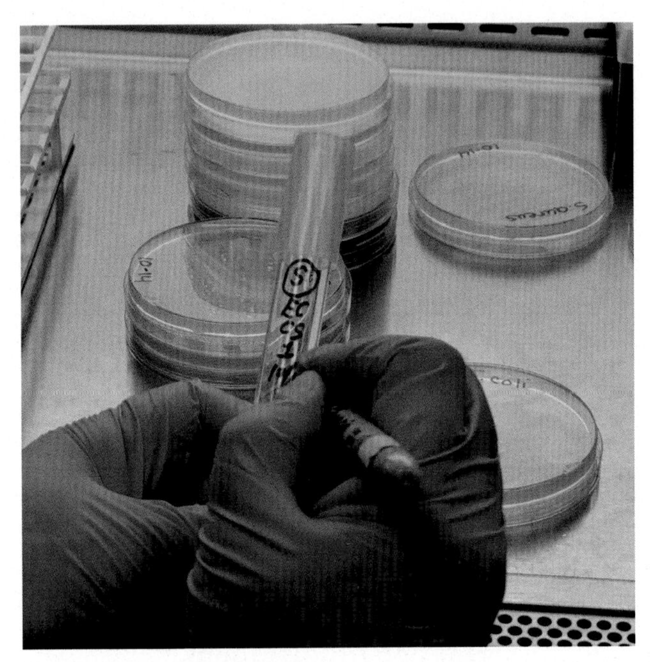

Fig. 15.4 ***Label the fermentation test tube.***

2. Incubate the test tube by using sterilized loop picking single colonies from the agar plates as shown in Figs. 15.5 and 15.6.
3. Incubate at 37°C for 24 h.

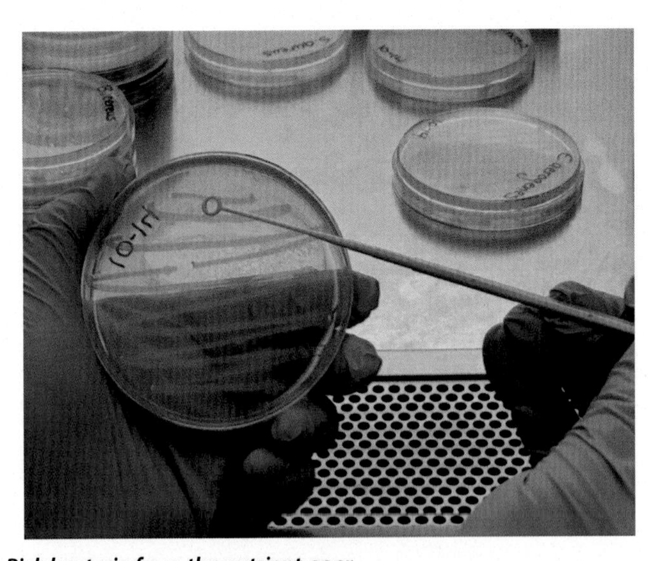

Fig. 15.5 ***Pick bacteria from the nutrient agar.***

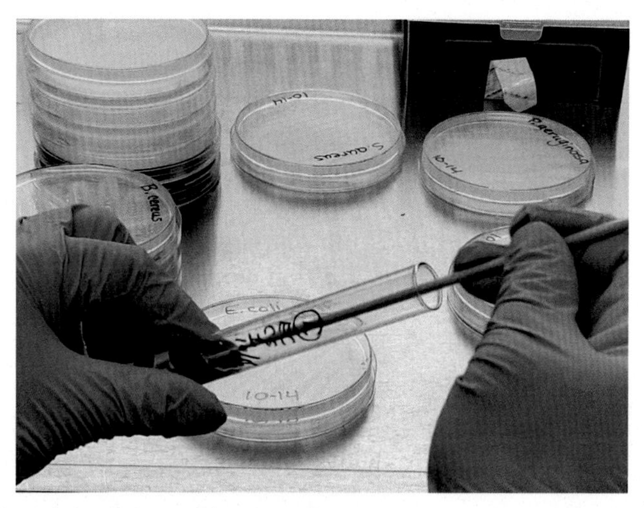

Fig. 15.6 ***Inoculate a fermentation test tube.***

4. Interpret results
   a. A positive culture will change the color of the medium from red to yellow (this indicates the production of acidic products as the result of fermentative metabolism).
   b. The presence of a bubble in the Durham tube indicates the production of gas during the fermentation process.
   c. A bright cherry red color is an indication of alkaline pH and indicates the organism did not ferment the sugar provided.
   d. Record whether your organism produces acid, gas, or alkaline by-products from the carbohydrate broth.

15.3 Exercise 3. Catalase test: One glass slide/student

15.3.1 Introduction

Some bacteria can reduce diatomic oxygen to hydrogen peroxide or superoxide. Both of these molecules are toxic. A defense mechanism which can minimize the harm done by the two compounds is the production of enzymes: superoxide dismutase and catalase. Both enzymes together will convert the super oxide back into diatomic oxygen and water. The catalase test involves adding hydrogen peroxide to a culture sample or agar slant. If the bacteria in question produce catalase, they will convert the hydrogen peroxide into water and oxygen gas. The evolution of gas, causing bubbles, is indicative of a positive test. This test is never to be done directly on blood agar plate medium.

Two methods for performing catalase test.

15.3.1.1 Method 1

1. Add several drops of hydrogen peroxide to the growth on your assigned plate cultures (we can use the spirit blue agar from lab section chapter 18). A positive reaction will be the production of oxygen gas bubbles as shown in Fig. 15.7.

Fig. 15.7 ***Catalase test with gas bubbles.***

15.3.1.2 Method 2

Procedure:

1. Place one drop of hydrogen peroxide on a clean glass slide.
2. Using your inoculating loop, pick an isolated colony from your assigned species and slowly immerse the cells into the drop of hydrogen peroxide (H_2O_2).
3. A positive test is read directly from the slide. Look for oxygen gas bubbles.

15.4 Exercise 4. Oxidase test

15.4.1 Introduction

Cytochrome oxidase is an enzyme found in some bacteria that transfers electrons to oxygen, the final electron acceptor in some electron transport chains. Thus, the enzyme oxidizes reduced cytochrome C to make this transfer of energy. Presence of cytochrome oxidase can be detected using oxidase test reagent (1% tetra methyl-para-phenylenediamine dihydrochloride). This reagent acts as an electron donor to cytochrome oxidase. If the bacteria oxidize the oxidase test reagent (remove electrons) the reagent will turn purple, indicating a positive test. No color change in 1 min indicates a negative test.

Procedure:

1. Label your plates as shown in Fig. 15.8.

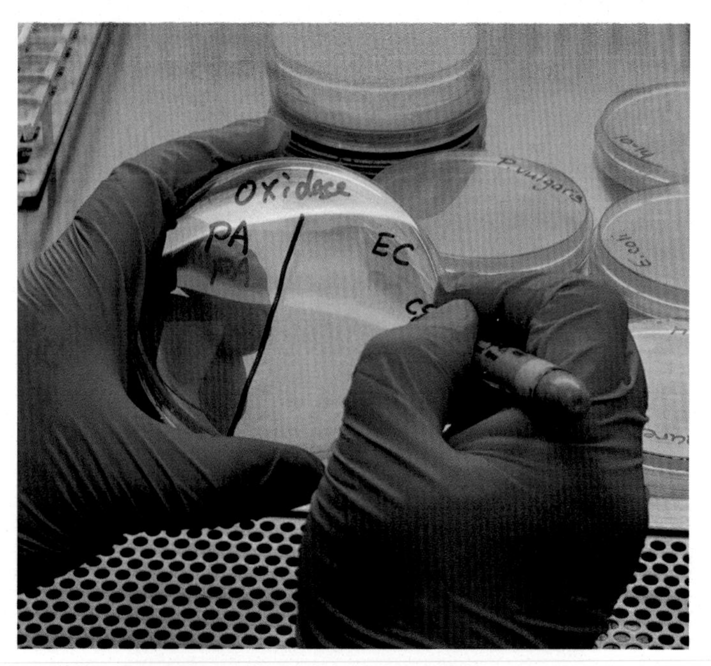

Fig. 15.8 ***Label the bottom of an oxidase test agar plate.***

2. Please split your agar into two sections, one section streak inoculates *Pseudomona aeruginosa*, the other section streak inoculates your assigned culture from following *Staphylococcus aureus*, *Escherichia coli*, *Bacillus subtilis*, *Proteus vulgaris*, *Enterobacter aerogenes* as shown in Fig. 15.9.

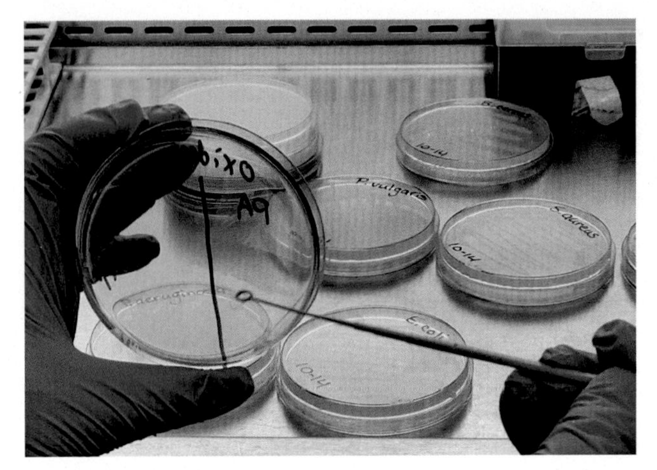

Fig. 15.9 ***Inoculate bacteria onto oxidase test agar.***

3. Incubate at 37°C for 24 h.
4. Add two to three drops of oxidase reagent (*p-aminodimethylaniline*) as shown in Fig. 15.10 to the surface of the growth. Positive results turn purple after 10–30 sec, negative results turn colorless as shown in Fig. 15.11.

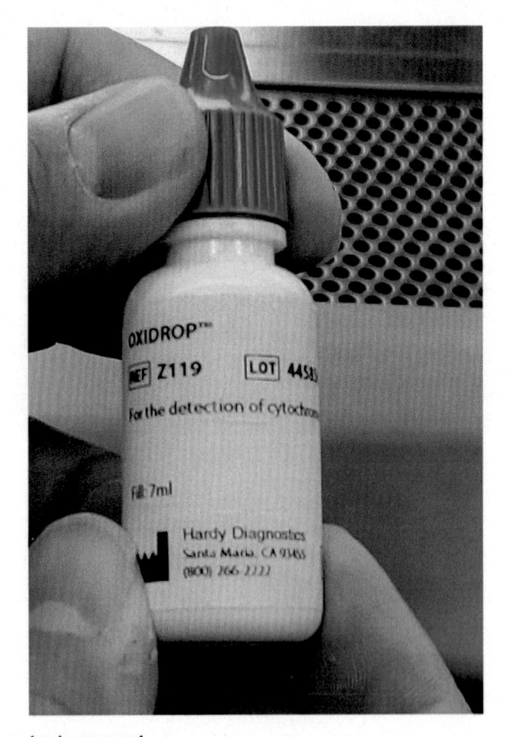

Fig. 15.10 ***Oxidase test reagent.***

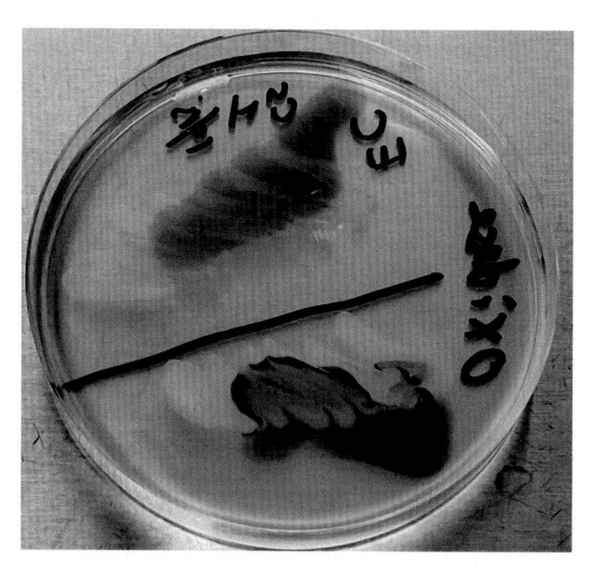

Fig. 15.11 ***Oxidase test results on the agar plate.***

15.5 After class questions

1. What can we say about those bacteria that produce alkaline by-products?
2. In clinical laboratory settings isolated pathogens are grown on 5% sheep blood agar plates. Why would it be improper to perform the catalase test on cells grown on blood agar?
3. Please explain the results of fermentation test as shown in Fig. 15.12.
4. Please explain the results of urea test as shown in Fig. 15.13.

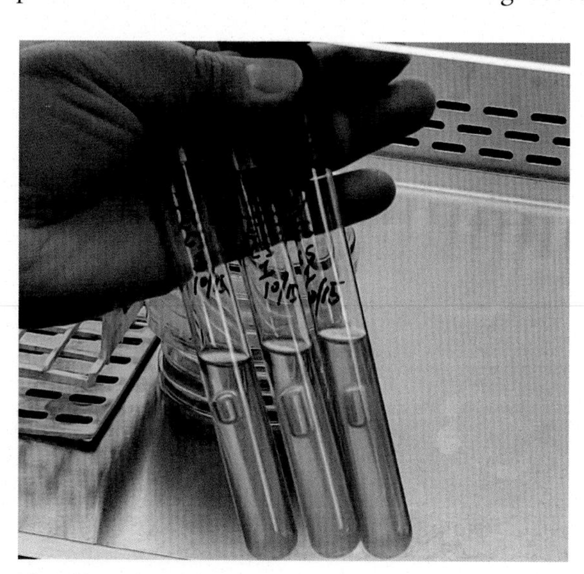

Fig. 15.12 ***Results of fermentation test.***

Fig. 15.13 ***Results of urea test.***

CHAPTER 16

Biochemistry test of bacteria-2 (nitrate reduction, decarboxylation, & deamination of amino acids, IMViC test, hydrogen sulfide, motility, and indole test)

Chapter outline

The following bacteria will be assigned to each student: *Staphylococcus aureus*, *Escherichia coli*, *Bacillus subtilis*, *Proteus vulgaris*, *Pseudomonas aeruginosa*, *Enterobacter aerogenes*.

16.1 Exercise 1. Sulfate indole motility test: One sulfate indole motility medium tube/student

16.1.1 Introduction

In this exercise you will use the inoculating needle and stab inoculate when you did for the motility test. The sulfate indole motility (SIM) tube will allow you to conduct three tests in one tube, including sulfate reduction, indole production, and motility.

Introductory Microbiology Lab Skills and Techniques in Food Science
DOI: https://doi.org/10.1016/B978-0-12-821678-1.00029-0

16.1.1.1 *Sulfate reduction*

H_2S is produced when sulfate salts or sulfur containing amino acids are metabolized by microorganisms. The media contains metallic ions such as lead, bismuth, or iron which will combine with any H_2S produced and form a **BLACK** insoluble metal sulfide precipitate. The process is anaerobic and will most likely occur at the bottom of your SIM tube inoculation.

16.1.1.2 *Motility*

SIM media is a semi solid agar media much like motility test media which you have already used in the past. If you do not have a positive H_2S reaction, then you can determine motility for your assigned species in the same manner as with motility test medium.

16.1.1.3 *Indole reaction*

Indole is the metabolic breakdown of the amino acid tryptophan. The production of Indole must be seen visually by adding **Kovac's** reagent to the culture tube.

16.1.2 Procedure

1. Label a SIM tube as shown in Fig. 16.1.

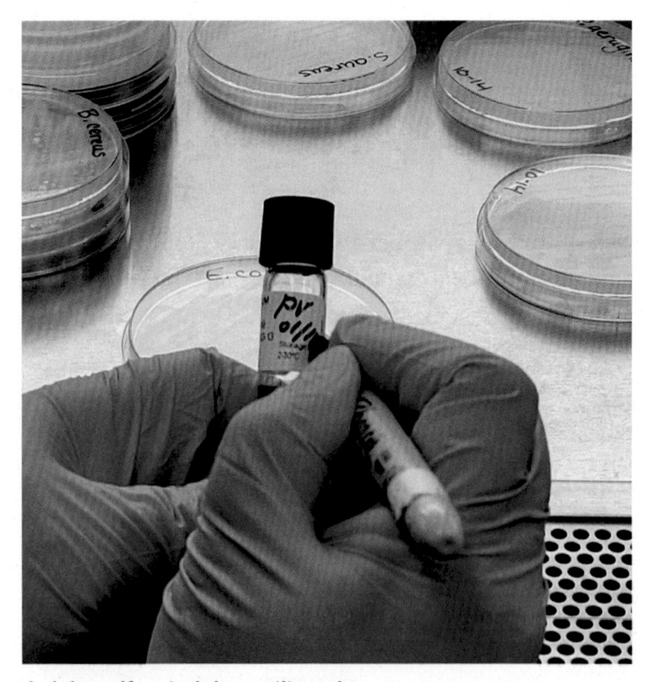

Fig. 16.1 ***Label the sulfate indole motility tube.***

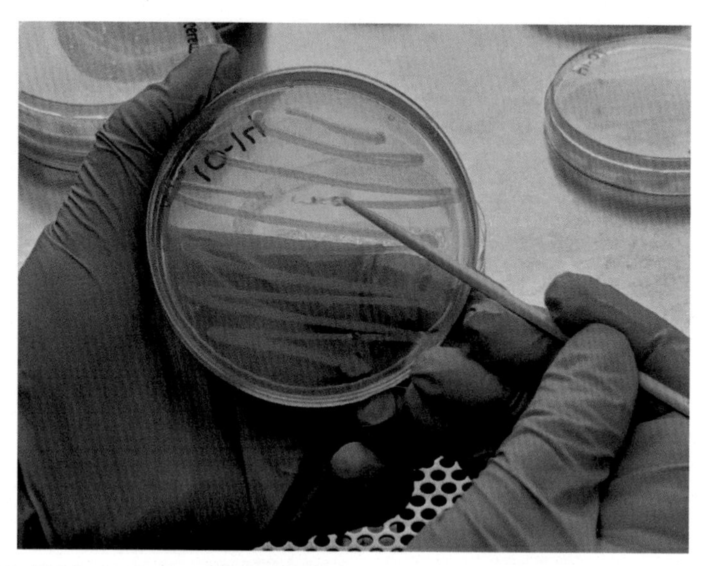

Fig. 16.2 ***Pick bacteria from the agar plate.***

2. Stab inoculate your assigned culture as shown in Fig. 16.2 to about two-third of the total volume of media as shown in Fig. 16.3, you will use a different species than your lab bench partner.
3. Incubate at 37°C for 24 h.
4. Examine the tube for evidence of hydrogen sulfide production.

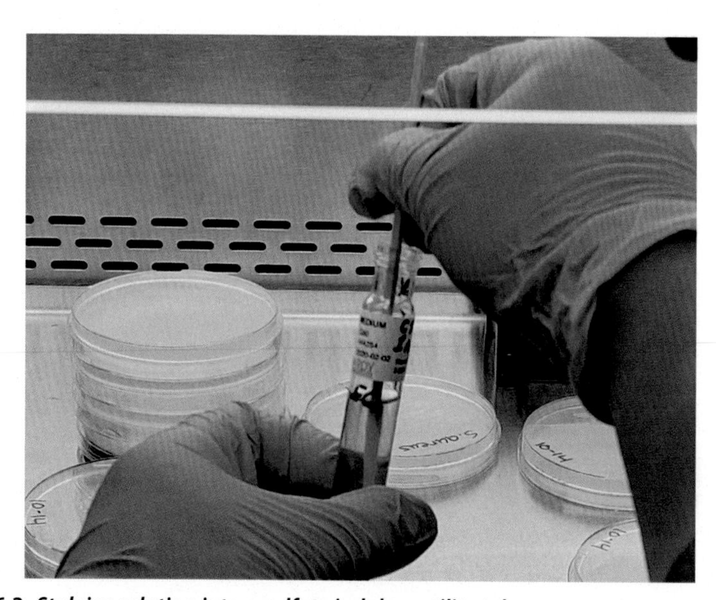

Fig. 16.3 ***Stab inoculation into a sulfate indole motility tube.***

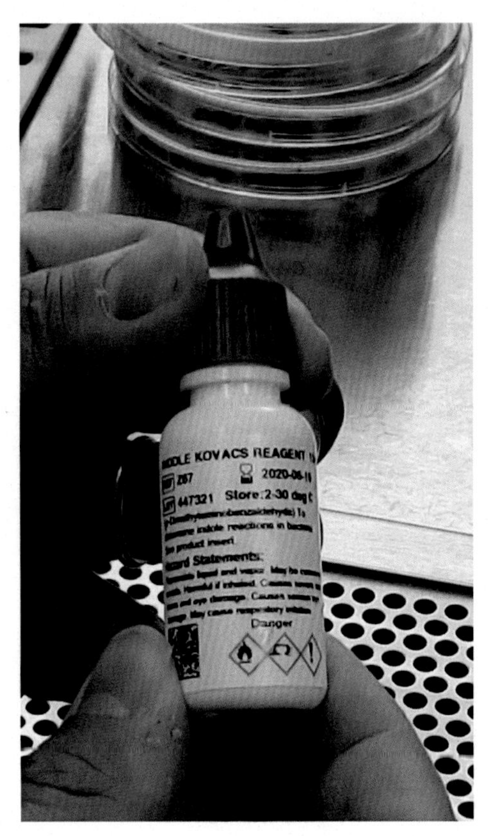

Fig. 16.4 ***Indole KOVAC reagent.***

5. Examine the stab line of inoculation for evidence of motility.
6. Using an Indole reagent dropper, add six or eight drops Kovac's Reagent (Fig. 16.4) to the SIM tube.
7. Allow the tube to sit for 1 min.
8. Observe the reagent for a "cherry red" color change. Cherry red is positive for the production of indole as shown in Fig. 16.5.

16.2 Exercise 2. Phenylalanine deaminase test

16.2.1 Introduction

Some bacteria produce enzymes that can metabolize amino acids. Before an amino acid can be used by the cell as an energy source the amino group must be removed, a process called deamination. In this experiment we will be using the medium phenylalanine agar. This medium contains the amino

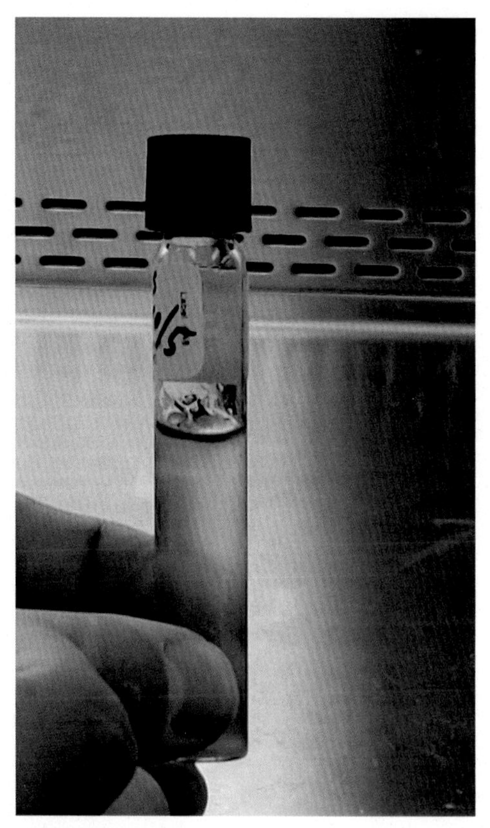

Fig. 16.5 ***Positive indole test result.***

acid phenylalanine. The enzyme phenylalanine deaminase will remove the amine group (NH_2) and release it as free ammonia (NH_3). This generates phenylpyruvic acid, which can be detected by adding an oxidizing reagent such as ferric chloride to the incubated tube. If the acid is present, a green color can be detected. The green color fades, so the test should be read immediately. A yellow color is a negative result.

16.2.2 Procedure

1. Label tube with your group number and assigned species as shown in Fig. 16.6.
2. Inoculate an isolated colony from your assigned culture onto one phenylalanine slant as shown in Fig. 16.7.
3. Incubate tube at 37°C for 24 h.

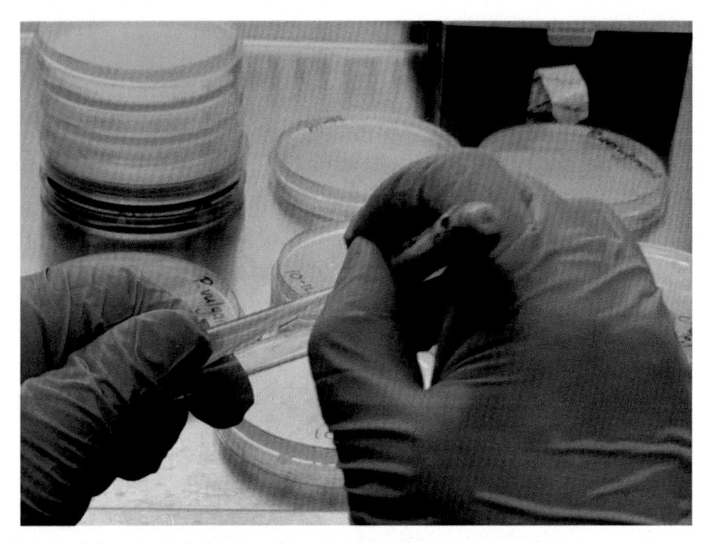

Fig. 16.6 ***Label the phenylalanine deaminase tube.***

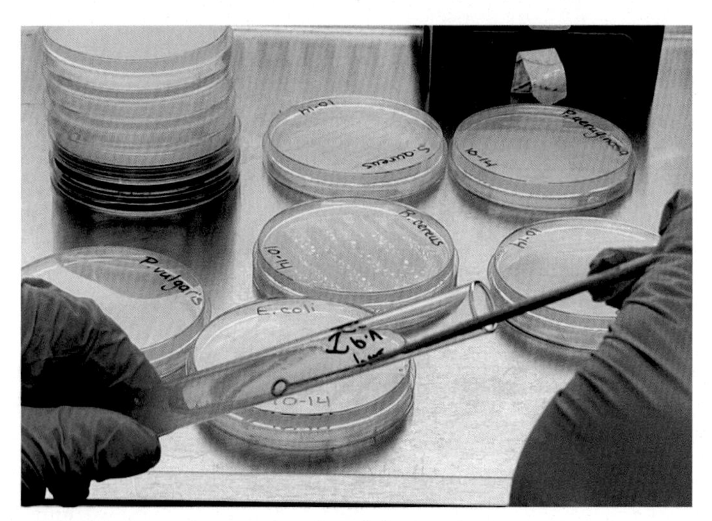

Fig. 16.7 ***Inoculate bacteria onto the phenylalanine slant.***

4. Test for the presence of phenylpyruvic acid.
   a. Using a reagent dropper, add 6 or 8 drops of 10% ferric chloride reagent to the phenylalanine agar slant as shown in Fig. 16.8.
   b. Allow tube to stay about 1 min.
   c. Observe slant for a dark green color that should develop within one minute as shown in Fig. 16.9.

Fig. 16.8 *Add ferric chloride reagent onto the phenylalanine slant.*

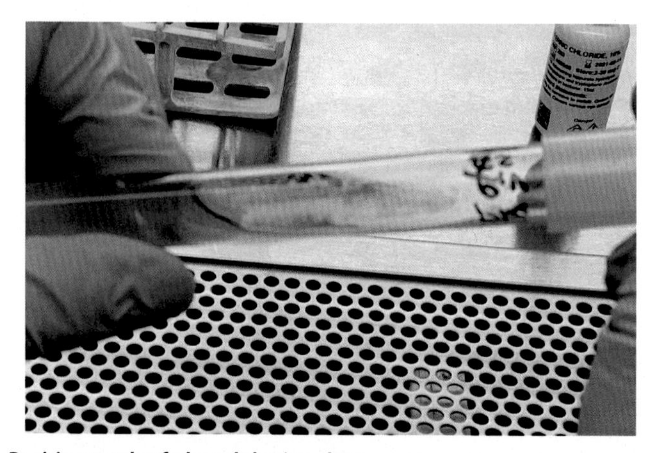

Fig. 16.9 *Positive result of phenylalanine slant.*

16.3 Exercise 3. Methyl Red Voges-Proskauer (MRVP) testing: Two MRVP tubes/group

16.3.1 Introduction

The methyl red test is used to determine the ability of an organism to produce and maintain stable acid end products from glucose fermentation, to overcome the buffering capacity of the system. If the organism can produce a variety of acids during the fermentation of glucose, you will see an immediate red color. The organisms that carry out mixed acid fermentation also generally produce gas (i.e., CO_2 and H_2). If a Durham tube is added to this media upon preparation in the laboratory, it may be used to detect this gas reaction.

The Voges-Proskauer test was designed for organisms that can ferment glucose but quickly convert their acid products to acetoin and 2,3-butanediol. If acetoin is present in the MRVP tube after 48 h. incubation the VP reagents will oxidize it to diacetyl resulting the RED positive test result. This test will indicate an organism's ability to produce neutral pH end products after fermenting the glucose.

16.3.2 Procedure

1. Your group will label and inoculate one MRVP tube with Escherichia coli (one tube labeled MR test and the other labeled VP test) as shown in Figs. 16.10 and 16.11.
2. Your group will label and inoculate one MRVP tube with *Enterobacter aerogenes* (one tube labeled MR test and the other labeled VP test).
3. Incubate your MRVP tubes 37°C for 24 h.
4. Add three to two drops of methyl red reagent to the each of the tubes labeled MR as shown in Fig. 16.12.

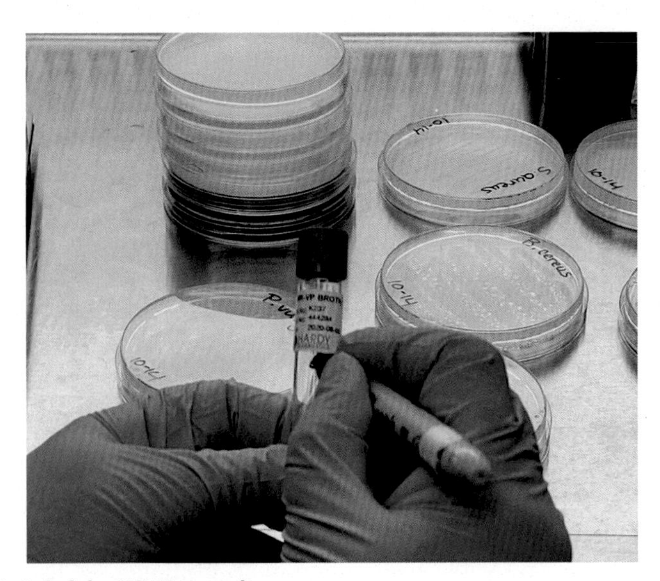

Fig. 16.10 ***Label the MRVP test tube.***

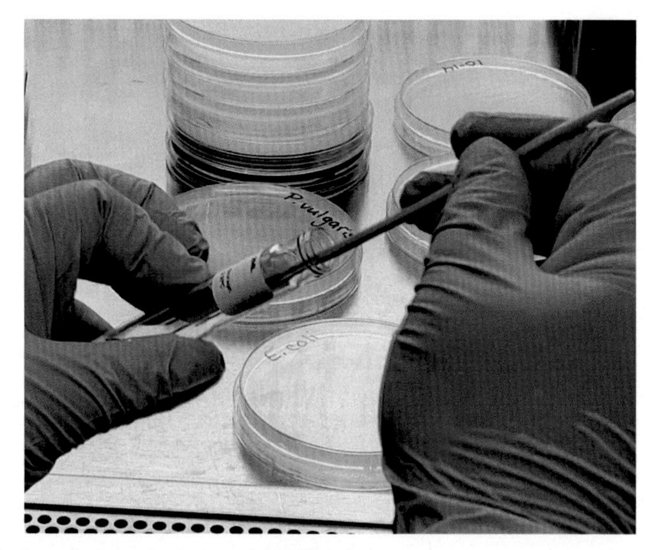

Fig. 16.11 ***Inoculate bacteria into a MRVP tube.***

Fig. 16.12 ***Add methyl red reagent into the MRVP test tube.***

5. Obtain and label two clean test tubes as shown in Fig. 16.13, one for each VP tube you have inoculated and incubated at 37°C for 48 h.
6. Using your automatic pipet pump and a sterile 15 mL pipet to transfer 1 mL of each 48 h VP tube culture to a clean, labeled, test tube as shown in Fig. 16.14.

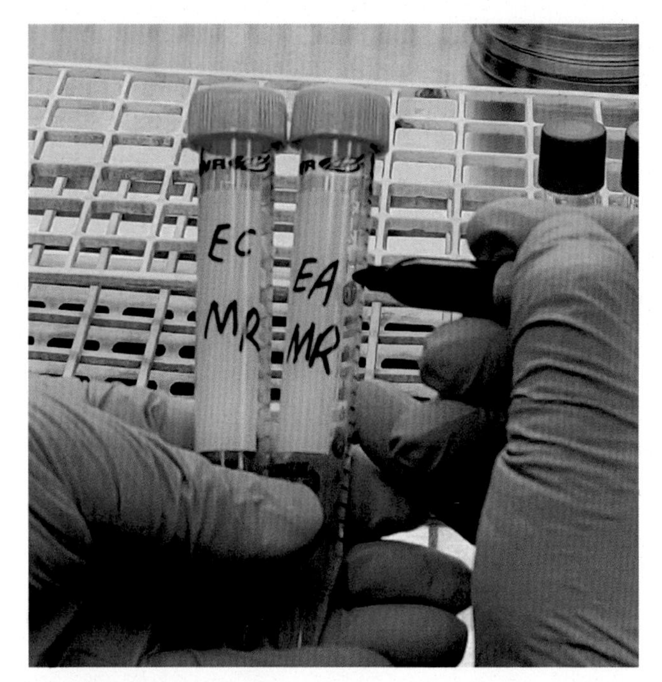

Fig. 16.13 ***Label the clean 15 mL tubes.***

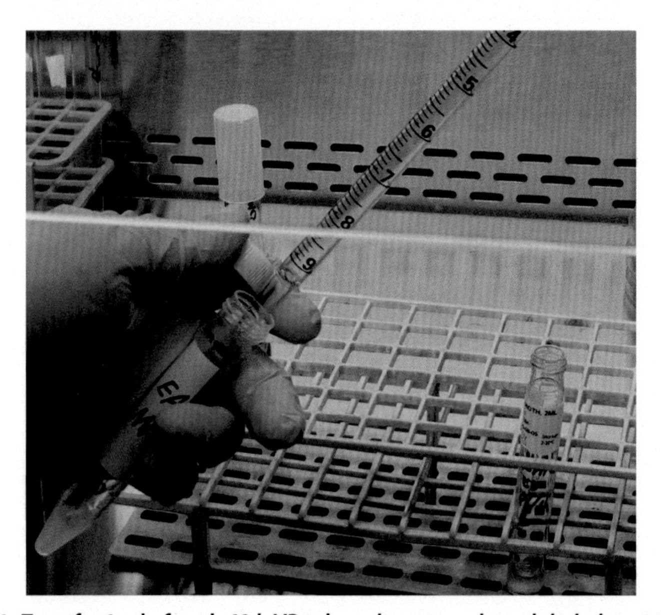

Fig. 16.14 ***Transfer 1 ml of each 48 h VP tube culture to a clean, labeled, test tube.***

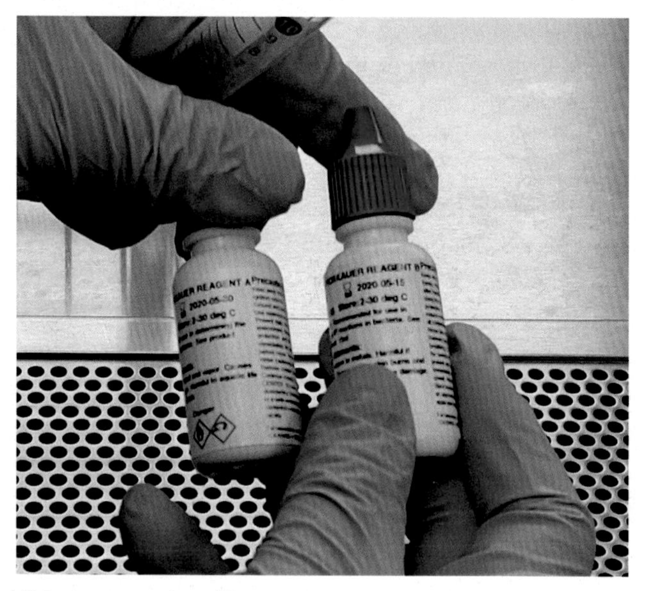

Fig. 16.15 ***VP test reagent A and B.***

7. Add six drops of VP solution A and two drops of VP solution B as shown in Fig. 16.15 to each test tube while shaking to aerate and mix the tubes.
8. Observer for 5 min. A positive VP test will turn Red indicating the presence of acetoin that was oxidized to diacetyl.

16.4 Exercise 4. Nitrate reduction test: One tube of nitrate broth/student

16.4.1 Introduction

The nitrate test is testing for the reduction of nitrate can be confusing. The key step is to detect nitrite in your broth culture (one of the possible end products of nitrate reduction).

16.4.2 Procedure

1. Label and Inoculate nitrate broth with your assigned species.
2. Incubated at 37°C for 48 h.
3. Examine the culture tube for growth.
4. Obtain and label a clean test tube.

5. Using your blue pipet pump and a sterile 5 mL pipet, you will transfer 2 mL of the 48-h nitrate broth culture to the clean test tube.
6. Add **five drops** of nitrate A solution and **five drops** of nitrate B solution to the 2 mL broth aliquot, shake and observe for a **red color** (positive) in 2 min. If negative the nitrate was either not reduced or it was reduced to a different end-product.
7. If negative test for the presence of **nitrate** that did not get reduced by dipping a wooden applicator stick into zinc dust and then using that stick to stir your reaction mixture from step 6.
8. Zinc dust will catalyze the reduction of any Nitrate to Nitrite. If a red color appears at this step the test is negative. If **no red color** appears the test is positive.

16.5 After class questions

1. Why MRVP test is important for differentiate *E. coli* and Enterobacter?
2. What indicates from a MR positive test result as shown in Fig. 16.16.
3. Please make your own flow chart of nitration reduction test.

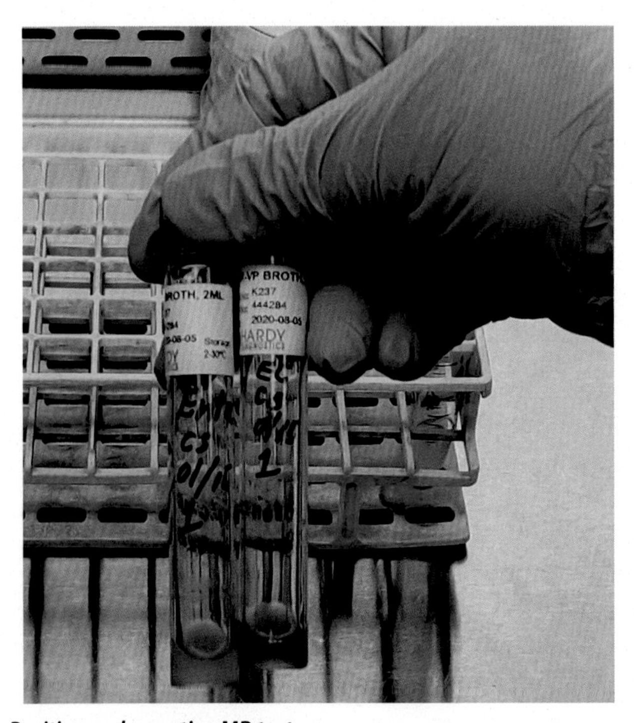

Fig. 16.16 ***Positive and negative MR test.***

CHAPTER 17

Biochemistry test of bacteria-3 (API20e and Enterotube II)

Chapter outline

17.1 Materials

Enterotube II, API-20e, *Escherichia coli*, *Salmonella*, MacConkey agar, XLT-4 agar

17.2 Introduction

We have practiced many individual biochemistry tests to identify certain bacteria. However, in real laboratory settings such as university laboratories, local clinics and reference, or small food quality control laboratories, it is impossible to identify bacteria correctly by using only a few tests, or through multiple extensive and expensive biochemical tests. Therefore, several commercially available multimedia system for bacterial identification have been developed and utilized successfully when speed as well as cost effective identification is needed. In this lab section, we will introduce and practice two most commonly used multimedia system Enterotube II and API-20e.

Enterotube II, as shown in Fig. 17.1, is a test kit looks like a pen with 12 components containing various biochemical substrates including carbohydrates such as glucose, lactose, adonitol, arabinose, sorbitol and dulcitol, H_2S and indole, carbinol, phenylalanine, urea, lysine and ornithine, and citrate. It can be inoculated easily from a single isolated colony on an agar plate. The Enterotube II is an example of a rapid multimedium test system used in identification of bacteria belonging to

Introductory Microbiology Lab Skills and Techniques in Food Science
DOI: https://doi.org/10.1016/B978-0-12-821678-1.00012-5

Enterobacteriaceae, which are characterized by non-endospore forming, gram negative, rod shape, facultative, lactose fermentation positive, and oxidase negative. Enterobacteriaceae normally present in the intestinal tract of humans and animals.

API-20e test, the API-20E test strip containing 20 separate test compartments is used to identify the enteric gram-negative rods shape bacteria. A bacterial suspension is added to each well which include dehydrated substrates. Some of the wells will have color changes due to pH differences: others produce end products that have to be identified with reagents. A profile number is determined from the sequence of positive and negative test results, then looked up in a code book having a correlation between numbers and bacterial species.

17.3 Exercise 1. Use Enterobute II to identify *Escherichia coli*

17.3.1 Procedure

1. Label the side of the Enterotube II as shown in Fig. 17.1.

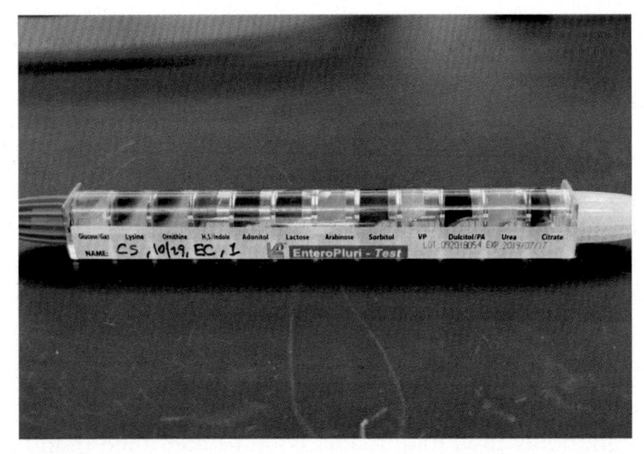

Fig. 17.1 ***Enterotube II.***

2. Remove the screw caps from both blue and white ends, aseptically pick one presumptive *E. coli* O157:H7 colony from MacConkey agar with the white sharp end as shown in Fig. 17.2.

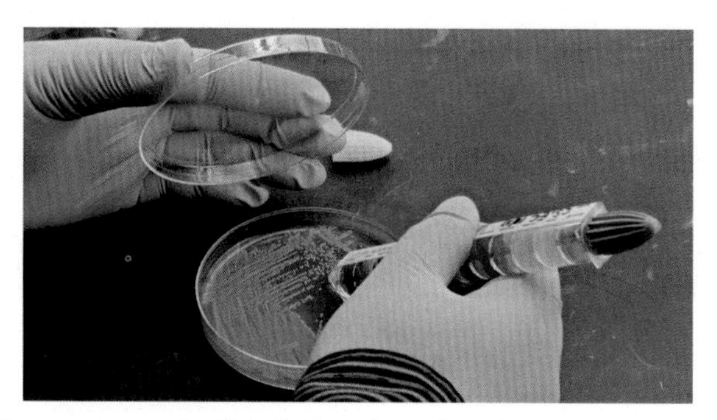

Fig. 17.2 ***Pick one presumptive*** **E. coli** ***O157:H7 colony from MacConkey agar.***

3. Pass the needle through the 12 wells and return it back to the original end, and screw the white cap back as shown in Fig. 17.3.

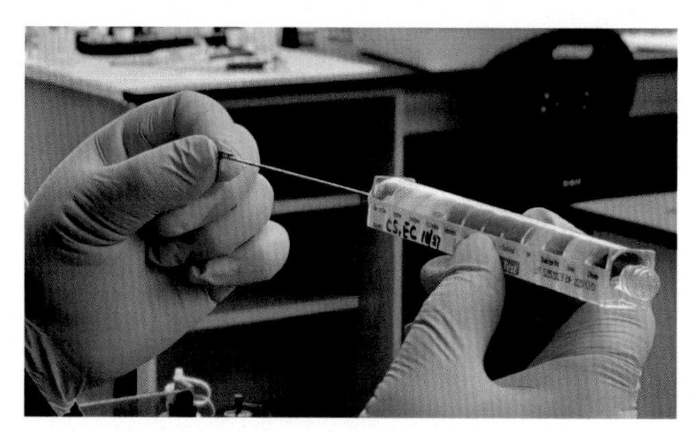

Fig. 17.3 ***Pass the needle through the 12 wells of Enterotube II.***

4. Bend and break the needle at the blue cap end as shown in Fig. 17.4, punch the plastic film of ADO, LAC, ARB, SOR, VP, DUL/PA, URE and CIT compartments as shown in Fig. 17.5. Screw the blue cap back.

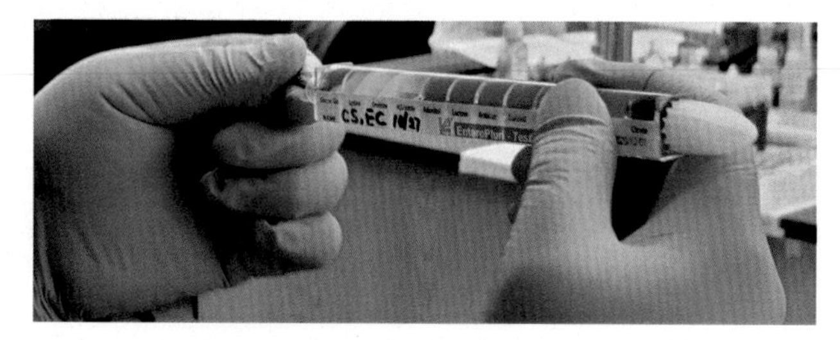

Fig. 17.4 ***Bend and break the needle at the blue cap end.***

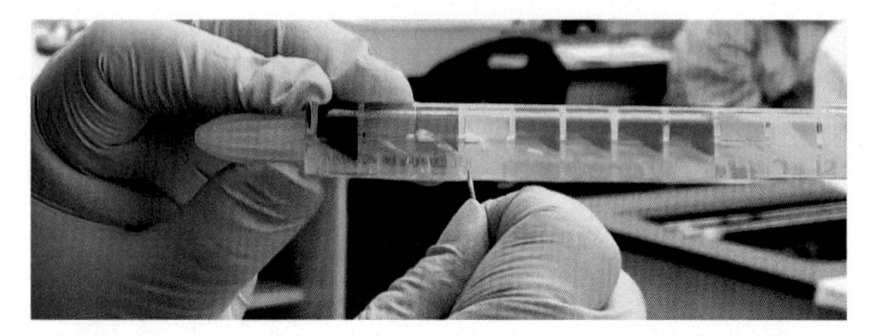

Fig. 17.5 ***Punch the plastic film.***

5. Incubate the Enterotube II at 35°C and read results in 24 h.
6. Interpret the results of your Enterotube II using the Table 17.1.

Table 17.1 Enterobute II positive and negative results table.

Well number	Substrate	Negative results	Positive results
1	Glucose fermentation	Red or orange	Yellow, circle the number 2 under glucose on results sheet
2	Gas production	Wax not lifted	Wax lifted, circle the number 1 under gas on results sheet
3	Lysine decarboxylase	Yellow	Purple, circle the number 4 under lysine on results sheet
4	Ornithine decarboxylase	Yellow	Purple, circle the number 2 under lysine on results sheet
5	H_2S production	Beige	Black, circle the number 1 under lysine on results sheet
6	Indole production	Need to conduct this test separately.	Need to conduct this test separately. If positive, circle the number 1 under lysine on results sheet
7	Adonitol fermentation	Red or orange	Yellow, circle the number 4 under adonitol on results sheet
8	Lactose fermentation	Red or orange	Yellow, circle the number 2 under lactose on results sheet
9	Arabinose fermentation	Red or orange	Yellow, circle the number 1 under arabinose on results sheet
10	Sorbitol fermentation	Red or orange	Yellow, circle the number 4 under sorbitol on results sheet

(continued)

Well number	Substrate	Negative results	Positive results
11	Voges-Praskauer test	Need to conduct this test separately.	Need to conduct this test separately. If positive, circle the number 2 under lysine on results sheet
12	Dulcitol fermentation	Any other color	Yellow, circle the number 1 under dulcitol on results sheet
13	PA deaminase	Any other color	Black or smoky gray, circle the number 4 under PA on results sheet
14	Urea hydrolysis	Beige	Red or purple, circle the number 2 under urea on results sheet
15	Citrate	Green	Blue, circle the number 1 under citrate on results sheet

7. Circle the numerical value of each positive reaction.
8. Add all the circled numbers in each bracketed section and enter the sum in the results sheet as shown in Fig. 17.6. You now have a 5-digit reference number.

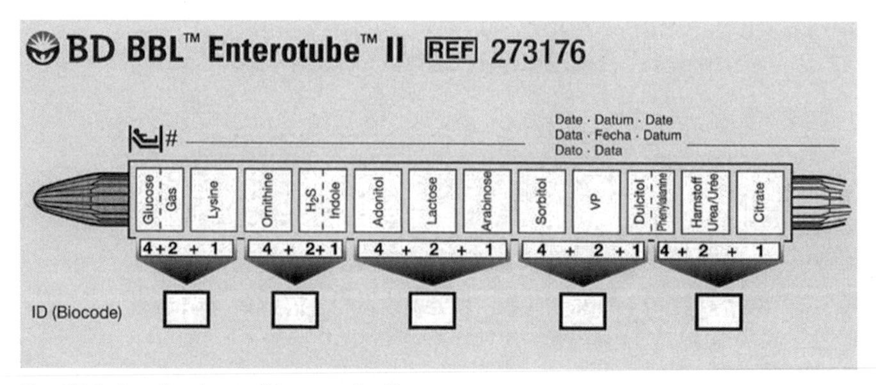

Fig. 17.6 ***Results sheet of Enterotube II.***

9. Locate the 5-digit number in the interpretation guide booklet and find the best identification in the column entitled "ID Value."
10. You may use the results sheet above to keep in your laboratory notebook. *E. coli* code should be 75340.

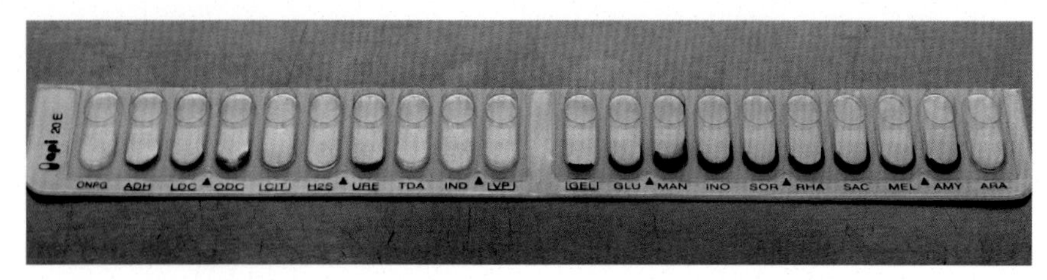

Fig. 17.7 ***API 20 test kit.***

17.4 Exercise 2. Use API 20e (Fig. 17.7) to identify *Salmonella* spp.

Note: This will be the extra exercise depending on the lab times of each semester.

1. Pick one *Salmonella* colony from prepared XLT-4 agar and suspend into a 5 mL of sterilized saline solution.
2. Use a bulb dropper to add the saline suspension into the cupules of the strip.
3. Fill the cupule with mineral oil for underlined test ADH, LDC, and URE.
4. Fill both the tube and cupule for boxed tests CIT, VP, and GEL.
5. Add 5 mL of water into the plastic base and place strip in the base.
6. Incubate the strip for 24 h at 35°C.

17.5 Validation after 24 h at 35°C

1. IND: Add one drop of Kovac reagent to record results in 2 min.
2. TDA: Add one drop of 10% ferric chloride immediately.
3. VP: Add one drop of 40% KOH and then one drop of 6% alpha-naphthol in 10 min.
4. Record the results (+ or -) on the worksheet as shown in Fig. 17.8.
5. Your instructor will interpret results of API-20e.
6. Record the API number and add number up for the positive reaction, and search the code in the API reference book.
7. Record and confirm the identity of the isolation in *Salmonella* spp.
8. A possible *Salmonella* spp. code is 6704752.

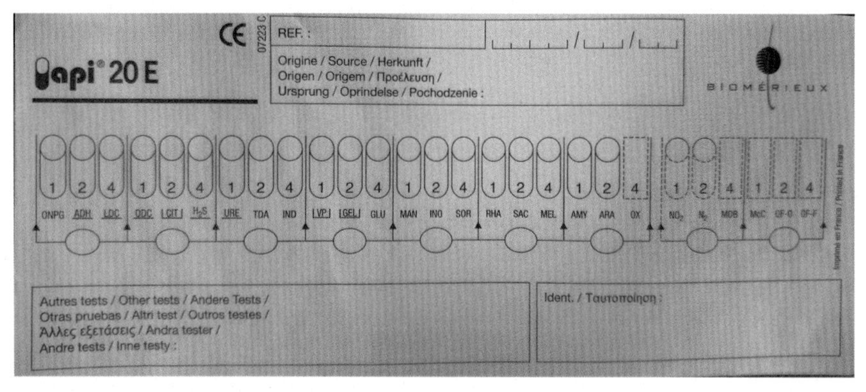

Fig. 17.8 ***API 20e results sheet.***

17.6 After class questions

1. What are enterobacteriaceae?
2. Why is it essential to have pure cultures for biochemical tests?
3. In the lab section, your instructor will interpret results of API-20e, please fill the result interpretation table as shown in Table 17.2 by yourself after class. One of your group members will present in front the classroom in the next lab section.

Table 17.2 Results interpretation of API 20e.

Well number	Test name	Negative results	Positive results
1	ONPG		
2	ADH		
3	LDC		
4	ODC		
5	CIT		
6	H_2S		
7	URE		
8	TDA		
9	IND		
10	VP		
11	GEL		
12	GLU		
13	MAN		
14	INO		
15	SOR		
16	RHA		
17	SAC		
18	MEL		
19	AMY		
20	ARA		
21	Oxidase test		

CHAPTER 18

Extracellular enzymatic activities of microorganisms

Chapter outline

18.1 Materials

Spirit blue agar, skim milk agar, starch agar, *Staphylococcus aureus*, *Escherichia coli*, *Bacillus cereus*

18.2 Exercise 1. Lipase test: One spirit blue agar plate/student

18.2.1 Introduction

The enzyme lipase will hydrolyze fats. Bacteria can be differentiated based on their ability to produce and secrete lipase. Many fats can be used in this assay but tributyrin oil is the most common. Spirit blue agar is prepared as an emulsion with tributyrin oil, but also contains spirit blue dye as a color indicator. The oil and dye will form a light blue complex in the media. Lipase positive bacteria will produce a clear halo effect around isolated colonies. Species of bacteria that produce lipase will demonstrate the hydrolysis of tributyrin with a clear halo surrounding the colony or growth. A species that produces a washed-out lightening of the media is not considered positive.

Introductory Microbiology Lab Skills and Techniques in Food Science
DOI: https://doi.org/10.1016/B978-0-12-821678-1.00002-2

18.2.2 Procedure

1. Label and divide a spirit blue agar plate into two sections as shown in Fig. 18.1.

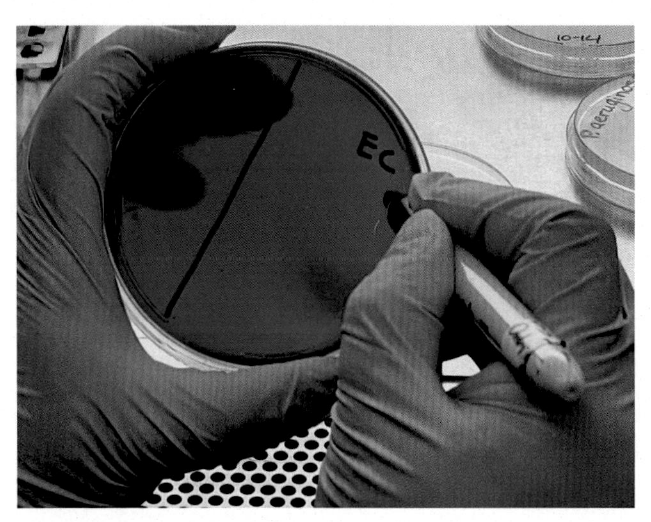

Fig. 18.1 ***Label the bottom of spirit blue agar.***

2. Inoculate *Staphylococcus aureus* on one side of your plate and *Escherichia coli* on the other side as shown in Fig. 18.2.

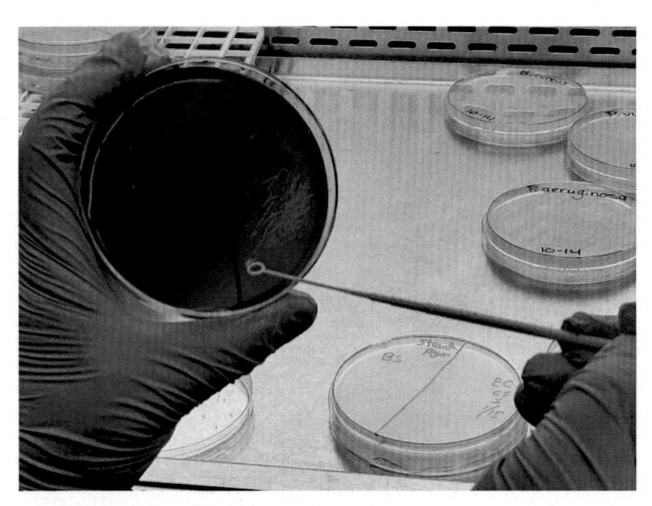

Fig. 18.2 ***Inoculation onto spirit blue agar.***

3. Incubate your plate at 37°C for 24 h.
4. Observe your spirit blue agar for the presence of lipase producing bacteria.

18.3 Exercise 2. Starch Hydrolysis: One starch agar plate/student

18.3.1 Introduction

Starch agar contains a polysaccharide made up of glucose subunits. Starch is too large to pass through the bacterial cell membrane, therefore, it must be hydrolyzed for it the have any metabolic value. Some bacteria produce two types of enzymes that are involved in the hydrolysis of starch: amylase and oligo-1,6, glucosidase. Amylase is for the hydrolysis of glucose in chains and oligo-1,6, glucosidase will hydrolyze the branching units of starch.

18.3.1.1 Starch hydrolysis test

Starch is soluble and virtually invisible in the medium therefore you must detect the presence or absence of starch using a starch indicator reagent. The reagent Iodine is used to detect the presence of starch by complexing with the molecule turning it blue to dark brown. If your starch agar medium turns blue to dark brown around your assigned bacterial species after the addition of Iodine you do not have starch hydrolysis.

18.3.2 Procedure

1. Label and divide a starch agar plate into two sections as show in Fig. 18.3.

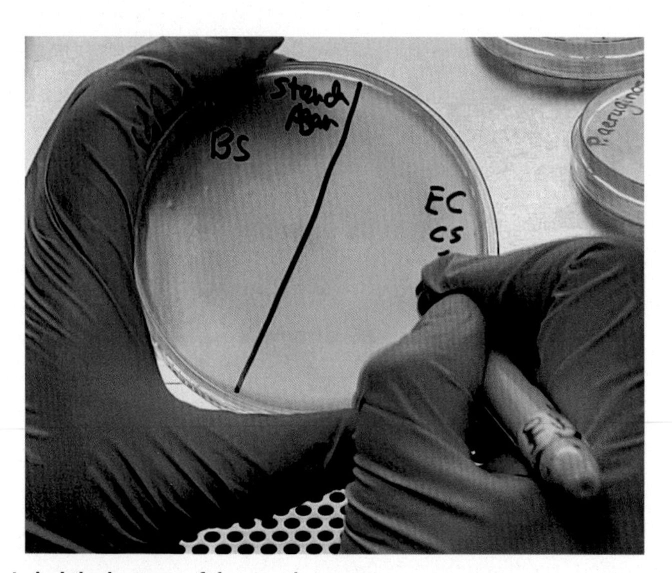

Fig. 18.3 ***Label the bottom of the starch agar.***

2. Inoculate *Bacillus cereus* on one side of your plate and *E. coli* on the other as shown in Fig. 18.4.

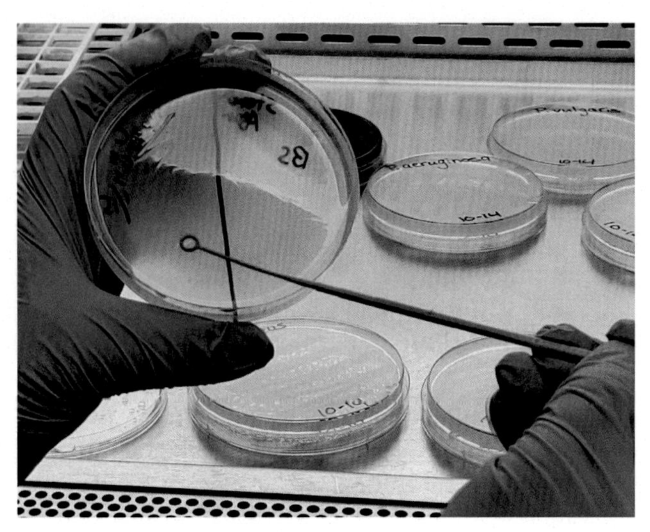

Fig. 18.4 ***Inoculation onto starch agar.***

3. Incubate your plate at 37°C for 24 h.
4. Remove the lid from your starch agar plate.
5. Flood the surface of the starch agar with iodine as shown in Fig. 18.5.

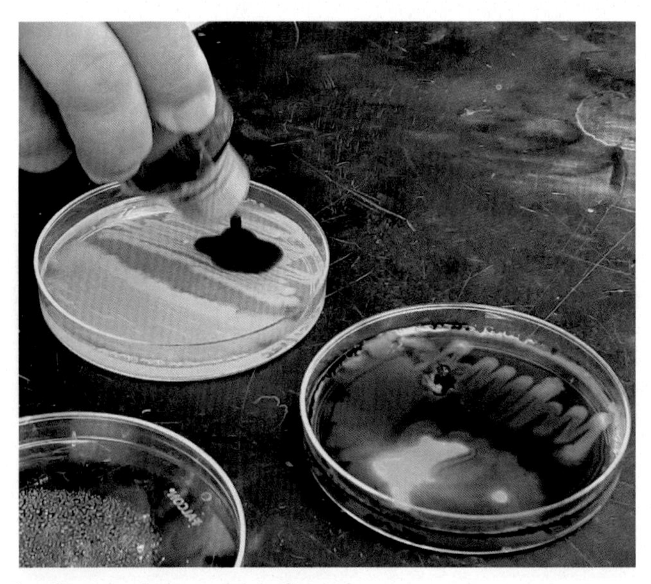

Fig. 18.5 ***Flooding starch agar with iodine.***

6. Observe your starch agar for the hydrolysis of starch (clear colorless zone around the assigned species is positive).

18.4 Exercise 3. Protein hydrolysis (protease test): One skim milk agar plate/student

18.4.1 Introduction

Many bacteria require proteins as a source of amino acids and other components for synthetic process. Proteins are biochemical compounds consisting of one or more polypeptides typically folded into a globular form, facilitating a biological function. One of the protein components in milk is called casein. The ability to break down this protein is called Casein Hydrolysis and is facilitated by the enzyme casease. Casein is an excellent protein for this hydrolysis assay due to the presence of the inorganic element calcium, which gives casein a white pigment and allows us to visually detect the protein.

18.4.1.1 Casein hydrolysis

If your assigned species produces casease, the enzyme will diffuse into the medium around the colony and create a zone of clearing (positive Hydrolysis). Bacterial species that cannot produce casease will not produce a clear zone around the growth, thus a negative test.

18.4.2 Procedure

1. Label and divide a skim milk agar plate into two sections as shown in Fig. 18.6.

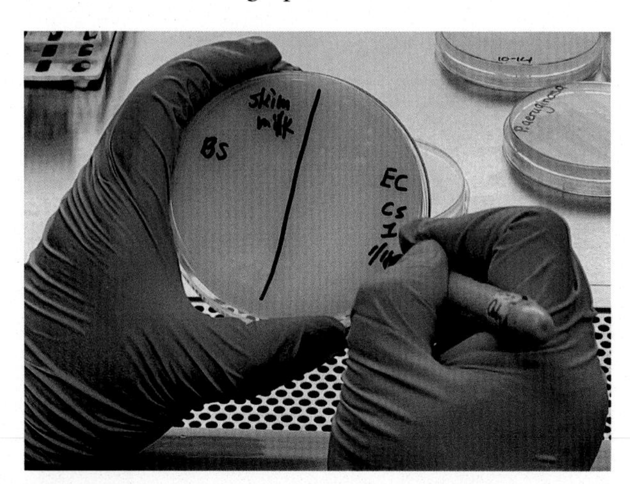

Fig. 18.6 ***Label the bottom of skim milk agar.***

2. Inoculate *Bacillus cereus* on one side of your plate and *E. coli* on the other side as shown in Fig. 18.7.
3. Incubate your plate at 37°C for 24 h.
4. Observe your skim milk agar for the hydrolysis of casein (clear zone around the assigned species is positive).

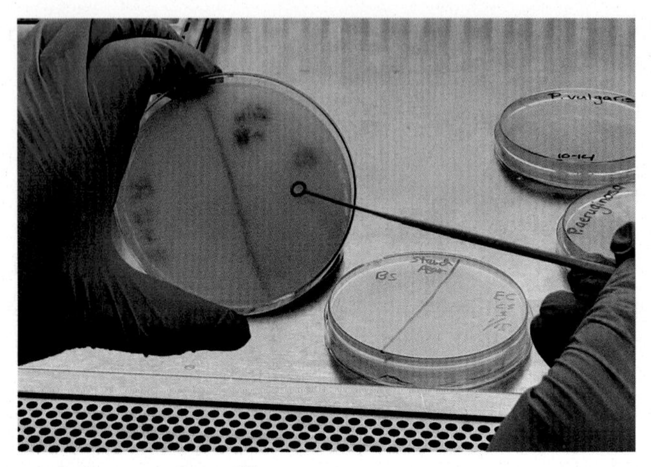

Fig. 18.7 ***Inoculation onto skim milk agar.***

18.5 After class questions

1. Describe a method of testing for starch hydrolysis and state how to interpret the results.
2. Describe a method of testing for lipids hydrolysis and state how to interpret the results.
3. Below are the results of skim milk agar as shown in Fig. 18.8, please explain what the results tell us.

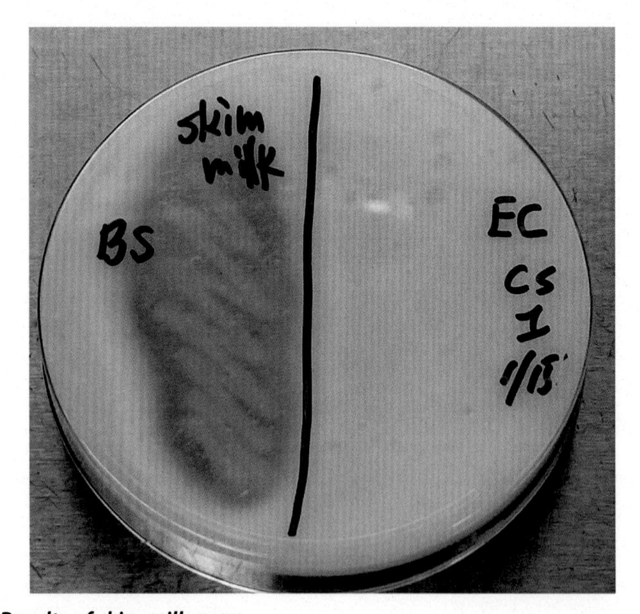

Fig. 18.8 ***Results of skim milk agar.***

CHAPTER 19

Case study 1 (bacteria cause upper respiratory tract diseases)

Chapter outline

19.1 Introduction

The human mouth is the reservoir of various microorganisms as its normal bacterial flora. The most dominant bacteria are the viridian streptococcus. In addition, numerous species of *Staphylococcus*, *Klebsiella pneumonia*, non-pathogenic Corynebacterium are also widely present in the oral cavity. While the presence of these normal bacteria protects us from certain disease, they can contribute to the development of dental caries and upper respiratory tract (URT) disease if they translocated, referred as opportunistic pathogens. Besides, *Streptococcus pyogenes* (group A Strep) existing in the tonsil area is responsible for disease in the throat, which can be differentiated in the sheep blood agar as compared to the other cultures. Below, we will briefly review the characteristics of several bacterial pathogens.

19.2 Lancefield groups

Lancefield, Rebecca Craighill (1895–1981), is an American microbiologist, introduced a new method to classify streptococci into groups based the ability of bacterial cell wall antigens to cause precipitation of the streptococci from solution after inducing the formation of their antibodies. These groups are now known as the Lancefield groups. The most important examples of streptococci classified within the Lancefield scheme in the clinical medicine are group A *S. pyogenes*, group B *S. agalactiae*, and group D *Enterococcus* spp.

Introductory Microbiology Lab Skills and Techniques in Food Science
DOI: https://doi.org/10.1016/B978-0-12-821678-1.00030-7

19.3 Streptococci

1. **α-Hemolytic streptococci:** *Streptococcus pneumonia* and *Viridian Streptococcus*, producing a dark green colony on blood agar due to the incomplete destruction of the red cells within the blood agar, reflecting the presence of biliverdin and other heme compounds. The distinction of pneumococci from other streptococci is optochin sensitivity. Typically, *S. pneumonia* form a 16-mm zone of inhibition around a 5-μg optochin disc (P-disk), and *V. Streptococcus* is resistant to optochin.
   a. ***S. pneumoniae***: Gram-positive, lancet-shaped cocci with the arrangement as pairs of cocci referred as diplococci, may also occur singly and in short chains. They are catalase test negative and ferment glucose to lactic acid. They do not belong to a Lancefield Group due to the lack of M protein in their cell wall. They are responsible for 80% community acquired pneumonia of adults in the U.S. They cause otitis for children between ages of 2 and 5 years with ear infection showing symptoms of ringing, buzzing, and ear pain. Since they contain heavy capsule surrounding the diplococci, they also show invasive effect with bacteremia and septicemia.
   b. ***V. Streptococcus***: The viridans streptococci normally inhabit the mouth and gastrointestinal tract and produce a greenish discoloration (viridans) on blood agar. The most notable viridans streptococci are *S. bovis*, *S. mitis*, *S. oralis*, *S. mutans*, *S. salivarius*, and *S. sanguis*. These organisms are leading cause of subacute bacterial endocarditis. They commonly cause catheter-related infections; and frequently are found in purulent abdominal, hepatobiliary, brain, and dental infections.
2. **β-Hemolytic streptococci:** *S. pyogenes* **(group A)** and *S. agalactia* **(group B)** producing transparent zone on the blood agar with destroying the red cells. Bacitracin A discs are used for differentiation of group A, beta hemolytic streptococci from other beta hemolytic streptococci. Bacitracin is active mainly against gram-positive organisms and inhibits cell wall synthesis of actively growing cells. Group A streptococci to be much more sensitive to bacitracin than other beta hemolytic streptococci, including group B streptococci.
   c. ***S. pyogenes***: It is Gram-positive, cocci in chains, and very invasive human pathogen causing strep throat and reserved in tonsil area. Clinical features include but not limited to pharyngitis, tonsillitis, sinusitis, otitis, arthritis, and bone infections. Some strains prefer skin, producing either superficial (impetigo) or deep (cellulitis) infections.

Table 19.1 The characteristics of *Streptococcus*.

Name	Gram-stain results	Hemolytic reaction on blood agar	Catalase test	Lancefield group	Identification
Streptococcus pneumoniae					
Viridans Strep					
Streptococcus pyogenes					
Streptococcus agalactiae					

People at all ages can become infected with this pathogen. Children at age 5–15 years have typical strep throat with strawberry tongue. People with pre-existing conditions like cancer, diabetes, and kidney disease, and those who use medications, such as steroids, are at high risk for this invasive disease. Breaks in the skin, like cuts, surgical wounds, or chickenpox blisters, can also provide an opportunity for the pathogen to enter the body. Sequelae for *S. pyogenes* may occur 1–3 weeks after acute disease include acute rheumatic fever and glomerulonephritis.

d. ***S. agalactiae***: It originally produce disease in cows. Pregnant women at 35–37 weeks are recommended to do a screening test of this group B strep of the vagina and rectum, otherwise could cause neonatal pneumonia, septicemia, and meningitis.

Exercise 1. Please summarize the characteristics of *Streptococcus* in the Table 19.1.

19.4 Staphylococci

Staphylococci are Gram-positive cocci, characteristically arranged in shape of grapes, and catalase test positive. They are non-endospore forming, facultatively anaerobic organisms that grow well on most nutrient media. They are the normal flora of our bodies reserved on skin and in the URT, including pharyngeal surfaces. There are three principal species: *Staphylococcus epidermidis*, *S. saprophyticus*, and *S. aureus*.

1. ***S. epidermidis***: They are the most frequent inhabitant of human surface tissues, including mucous membranes. It is not usually pathogenic, but it may cause serious infections if it has an unusual opportunity for entry, such as in cardiac surgery or in patients with indwelling intravenous catheters who usually have low resistance. They do not ferment

mannitol in mannitol salt agars, negative in coagulase test, and sensitive to low dose of novobiocin.

2. ***S. saprophyticus*:** They are referred as "honeymoon bacteria" and has been implicated in acute urinary tract infections in young, sexually active women, ages 16–25 years. It has not been found among the normal flora and in not yet known to cause other types of infection. They do not ferment mannitol in mannitol salt agars, negative in coagulase test, but resistant to low dose of novobiocin.
3. ***S. aureus*:** They produce toxins and enzymes that can exert harmful effects of the infected host. Their hemolysins can destroy red blood cells, coagulase coagulates plasma, leucocidin destroys leukocytes, and hyaluronidase brakes connective tissue. They mainly involved in three mechanisms of human disease: (1) Food intoxication: the enterotoxin is elaborated by *S. aureus* as they multiply in contaminated food such as chicken and egg salads. The enterotoxin is responsible for severe gastroenteritis or staphylococcal food intoxication. Symptoms including nausea, vomiting, cramping and diarrhea usually develop within 6 h after eating contaminated foods. (2) Toxic shock: *S. aureus* produce toxic shock syndrome (TSS) by elaborating a toxin referred to a TSST-1. This disease was seen primarily in menstruating women who use highly absorbent tampons back in 1970s. *S. aureus* colonizing the vaginal tract multiplies and releases TSST-1 cause a variety of symptoms including fever, rash, hypotension and shock. (3) MRSA: It is referred as methicillin-resistant *S. aureus* and is the most commonly pos-surgical nosocomial infection, which caused by widely used methicillin in hospital environment and careless of post-surgical hygiene. They ferment mannitol in mannitol salt agars, positive in coagulase test, sensitive to low dose of novobiocin, and showing β-hemolytic on blood agars.

Exercise 2. Please summarize the characteristics of *Staphylococcus* in the Table 19.2.

Table 19.2 The characteristics of *Staphylococcus*.

Name	Gram-stain results	Hemolytic reaction	Catalase test	Mannitol salt agar	Coagulase test	Reaction to novobiocin
Staphylococcus epidermidis						
Staphylococcus saprophyticus						
Staphylococcus aureus						

Exercise 3. Please read the cases below and recognize the pathogens related to the case. You can discuss with your lab bench partner and record your results in your lab notebook and explain the reason why this pathogen is chosen for the case.

Case 1. John, an English instructor at the university, received a call from the adult day care where his 74-year-old mother was staying. He was experiencing chest pain, chills, and had a fever. When he arrived at the day care she seemed to be short of breath and the cough she had that morning had worsened and became productive (meaning she is coughing up sputum). The local hospital prepared a sputum sample, which grew on blood agar and chocolate agar producing alpha hemolytic zones. The Gram stain revealed numerous white blood cells and numerous gram-positive diplococci that were catalase negative. An optochin susceptibility test was conducted, and it was sensitive to it.

Case 2. On Monday January 29, a 27-year-old female office assistant experienced a sudden onset of dizziness while at work. The lady was carried to a co-worker's car and began to vomit complaining of a severe headache, muscle aches and nausea. The lady was taken to the hospital where her condition worsened. She claimed she had not been feeling well all morning and presented with a temperature of 103°F. She developed a rash and experienced a sudden drop in blood pressure. Blood and CSF were negative for sepsis and bacterial meningitis. Her history was significant in that she was experimenting with a new contraceptive sponge that had not been removed since the previous evening. What organism is most likely causing this disease?

Case 3. A 22-year-old sexually active female comes into the student health clinic complaining of painful and frequent urination (dysuria). The technician spun down her urine for microscopic observation and found gram-positive cocci in grapelike clusters. Further testing revealed the organism was catalase positive and resistant to novobiocin.

Case 4. 29-year-old pregnant female elected to have home birth using a midwife. She chose to have little medical intervention and delivered a healthy 7 lb baby boy. On day 5 postnatal, the baby was experiencing tremors, vomiting, and tachycardia (abnormally rapid beating of the heart). He refused to breast feed and had a fever of 102°F. His urine, cerebral spinal fluid, and blood were cultured. The blood culture and CSF came back positive for gram-positive, catalase negative, beta hemolytic cocci. Considering the age of the patient which pathogen is most likely the cause of this sepsis. The isolate was resistant to bacitracin.

Case 5. A 2-year-old girl is admitted to the hospital with massive tissue destruction along her right arm. The skin is a violet color and large fluid-filled blisters are present. The patient has a fever, a rapid heart rate, low blood pressure, and seems confused. Her mother informs the physician that the child had been recovering from chickenpox, and, for the past two days, had frequently been scratching at chickenpox lesions on that area of her arm. Once the area appeared to have become infected, the infection spread very rapidly. A Gram stain of exudate from the infected tissue reveals gram-positive cocci in chains.

Case 6. An 80-year-old female is transferred from a nursing home to the hospital because she is suspected of having pneumonia. She is experiencing chest pain, chills, fever, and shortness of breath. She has a productive cough (meaning that she is coughing up sputum). A Gram stain of the sputum reveals numerous white blood cells and numerous gram-positive diplococci. Upon receipt of the Gram-stain report, the physician treats the patient for a pneumonia caused by which pathogen?

Exercise 4. Identification of normal bacteria in the throat.

This exercise is collaborated between you and your bench partner. One of you is presumptive to be a nurse, the other person is presumptive to be a patient with mild cold like symptoms.

19.5 Procedure

1. Obtain a blood agar and an autoclaved sterile cotton swab from the front desk.
2. Label the back of your blood agar with your patient's name, bench number, and date.
3. Have your "patient" open his/her mouth widely and saying "Ahhh" to flatten the tongue. Use the swab to gentle and firmly swab the tonsil area. Please avoid touching other area of the oral cavity including teeth, tongue, and cheek area.
4. Gently roll the swab on the first quadrant of the blood, and then switch to the sterile loop to complete the rest part as we did for streak plating.
5. Place the plate into a candle jar. The instructor will appoint one student from the classroom to flame the candle inside the jar and cover the lid to turn off the flame. This will create microaerophilic environment in the candle jar.
6. Incubate the candle jar at 37°C for 24 h.

7. After incubation, check the plate for the presence of pinpoint colonies with α-, β-hemolytic.
8. Conduct Gram-staining and catalase test for the picked pinpoint colonies.
9. Fill the results in Table 19.3 and complete the report including title, introduction, materials and methods, results, and discussion.

Table 19.3 Results table of throat culture.

Gram-stain reaction			
Hemolytic reaction and colony morphology Catalase test Suggested species confirmation test			

CHAPTER 20

Clinical lab practice - urine sample bacteria isolation and numeration

Chapter outline

20.1 Introduction

Normal urine should be sterile without containing any microorganisms. A urinary tract infection (UTI) is typically with bacteria and white blood cells in urines. Bacteria causing UTI are usually the patients' own normal intestinal microorganisms passing from the area near the anus into the urethra and ascend into bladder. The three major ways of people acquiring microorganisms in the urinary tract are (1) holding on the pee too long causing migration of fecal flora to the urethra, (2) catheterization with old catheter within patient for 1-week long without changing new ones, and (3) microorganism reach the kidneys via the blood stream. The most typical symptoms of UTI are called dysuria (difficult to pee with burning sensation on passing urine). The number one bacteria causing UTI is *Escherichia coli* (80% of UTI, may produce β-hemolytic zone), followed by *Klebsiella*, *Enterobacter*, *Serratia*, and *Proteus*.

20.2 Exercise 1. Dip stick urine test

The UTI can be presumptively diagnosed by dip-stick testing of urine samples through causal catch urine samples. Patient just pees into the urine bottle with dip-stick testing showing the presence of either or all of nitrites, white blood cells (leukocyte esterase) and blood.

Introductory Microbiology Lab Skills and Techniques in Food Science
DOI: https://doi.org/10.1016/B978-0-12-821678-1.00015-0

Nitrites: This test is based on the reaction of p-arsanilic acid and nitrite in urine to form a diazonium compound, which couples with N-ethylenediamine in acid medium and showing pink color.

White blood cells: This test is testing for leukocyte esterase, which cleave a derivatized pyrazole amino acid ester to release hydroxy pyrazole. The pyrazole reacts with a diazonium salt to produce a purple color. The detect limit is 15 white blood cells/mL.

Blood cells: Blood cells do not normally present in the urine. The presence of red blood cells may be associated with sever UTIs, kidney stones, and hemolytic anemia.

20.2.1 Procedure

1. Obtaining a simulated urine sample on the front desk as shown in Fig. 20.1, Or obtaining a new medical urine cup to get the real urine samples by yourself at restroom.

Fig. 20.1 ***Urine sample and test strip.***

2. Dip the test trip into the urine samples for 30 sec and remove it as shown in Fig. 20.2.

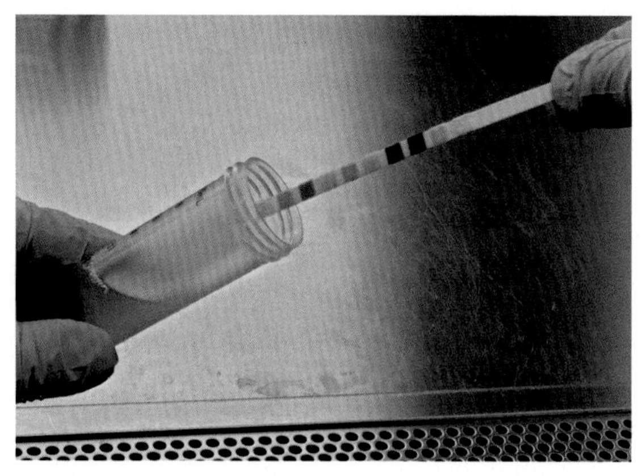

Fig. 20.2 ***Dip test strip into urine sample.***

Fig. 20.2. Dip test strip into urine sample

1. Please do not shake it to prevent cross contamination into the lab bench.
2. Read the strip by comparing it with the color chat on each bench, which is usually supplied by the manufacturers or on the side of the strip bottle as shown in Fig. 20.3.

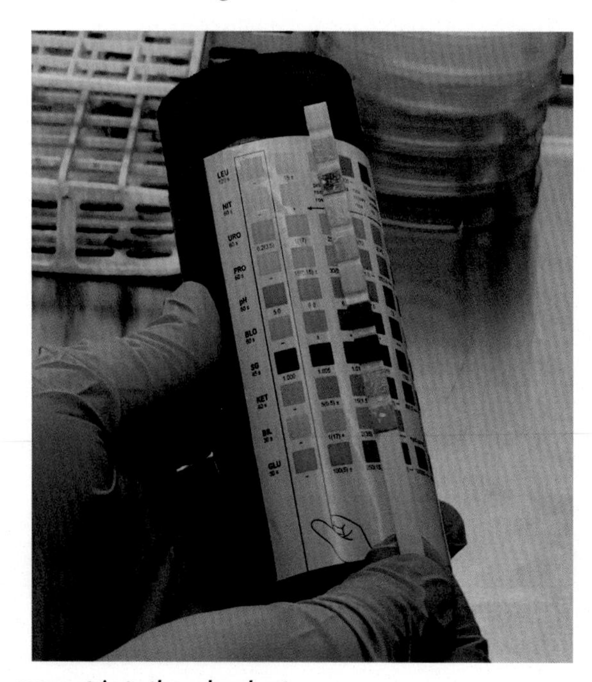

Fig. 20.3 ***Compare strip to the color chart.***

Table 20.1 Results of dip-stick test.

Test items	Dip test (positive or negative)
White blood cells	
Nitrites	
Blood	

3. Record your results in the Table 20.1.
4. Discard your strip into the biohazard trash can.

20.3 Exercise 2. Qualitative test of urine samples

This test is for testing the presence/absence of bacterial cells from urine samples on MacConkey agars and blood agars. A typical *E. coli* cells on MacConkey agar is pink colony showing lactose fermentation positive. A β-hemolytic zone shown on the blood agar indicate possible invasive infection from kidney.

20.3.1 Procedure

1. Obtaining one blood agar and one MacConkey agar from the front desk.
2. Label the bottoms of the agars as "qualitative test" together with your name initial, date, bench number.
3. Use the same urine samples from the dip-stick test to steak plating onto blood as shown in Fig. 20.4 and MacConkey agars as shown in Fig. 20.5 using sterilized plastic loop.

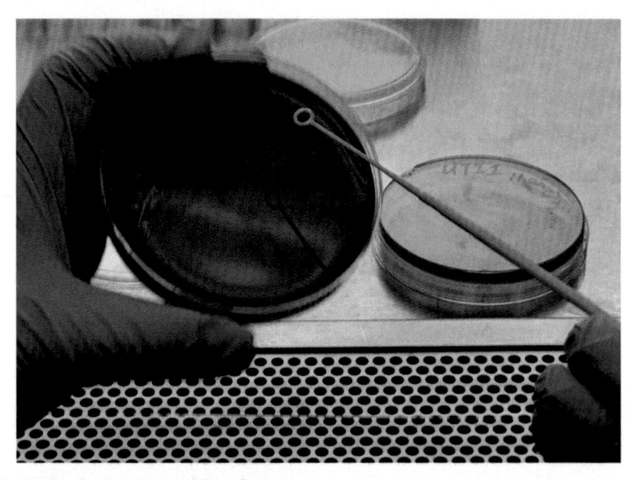

Fig. 20.4 ***Streak plating onto blood agar.***

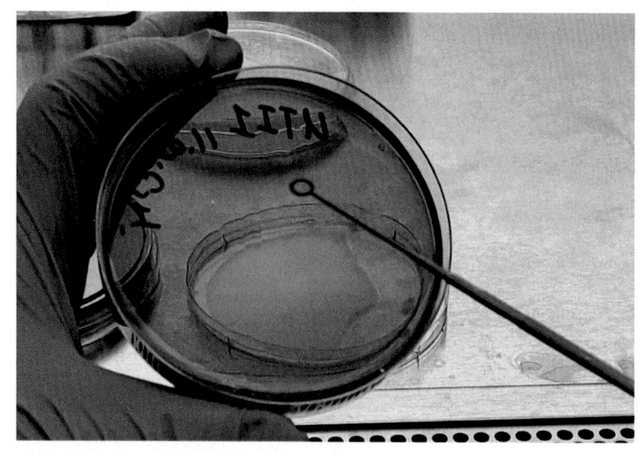

Fig. 20.5 ***Streak plating onto MacConkey agar.***

4. Incubate your plates at 35°C for 24–48 h.
5. Record results in Table 20.2.

Table 20.2 Results of qualitative test.

Agars	Growth or not, record if β-hemolytic on blood agars
MacConkey agar	
Blood agar	

20.4 Exercise 3. Quantitative test of urine samples

The clinical standard of degerming UTI is the bacterial cell numbers in the excess of 100,000 cells/mL of urine samples. Use sterile loop that can carry 0.001 mL of urine samples on the top of the loop as shown in Fig. 20.6.

Therefore, the final dilution factor is 1000 (1/0.001). The final bacterial cell concentration can be calculated by colonies of the plate multiple by 1000.

20.4.1 Procedure

1. Obtaining one nutrient agar from the front desk.
2. Label the bottoms of the agars as "quantitative test" together with your name initial, date, and bench number.

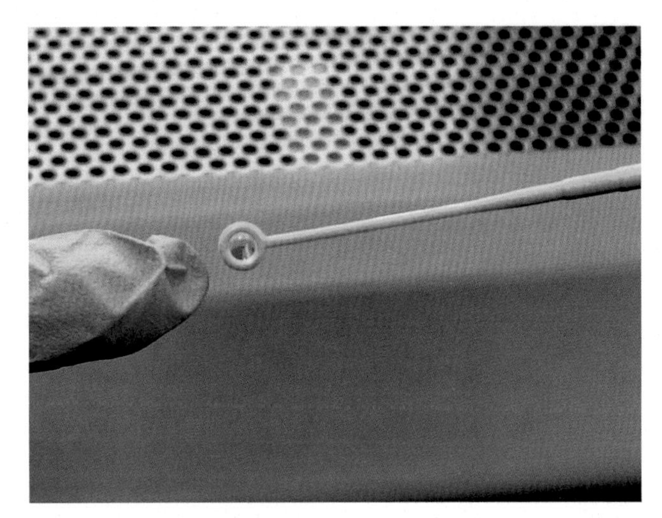

Fig. 20.6 ***Liquid carrying on the loop.***

3. Use the sterilized loop to gently streak the simulated urine samples onto the nutrient agars as shown in Fig. 20.7.

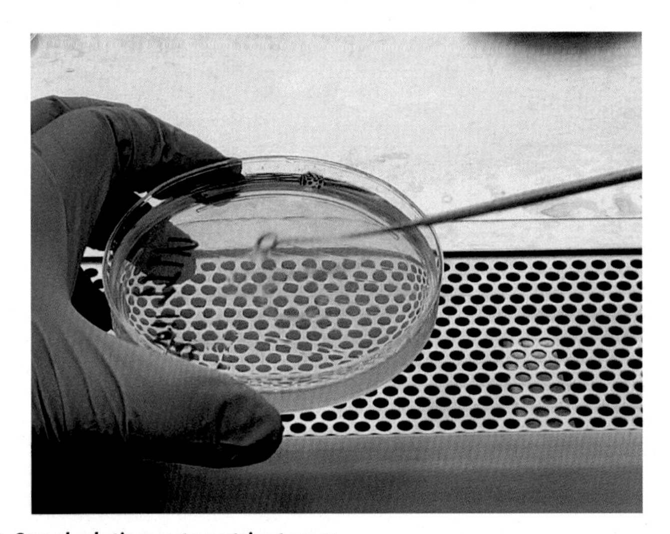

Fig. 20.7 ***Streak plating onto nutrient agar.***

4. Turn 90° angle and gently steak on the whole agar surfaces again.
5. Incubate your plates at 35°C for 24–48 h.
6. Calculate and record results in Table 20.3.

Table 20.3 Results of quantitative test.

Test items	Quantitative test
Number counted	
Concentration (CFU/mL)	
Significant for urinary tract infection (>100,000 CFU/mL)	

20.5 After class questions

1. How do microorganisms enter the urinary tract?
2. What do we learn from the dip-stick test, are the test results conclusive?
3. What can we conclude if a small pinpoint, beta hemolytic colonies on blood agar with no growth on MacConkey agar when conducting a qualitative urine test?
4. If a quantitative urine sample from a patient have 120 colonies, is this determined as UTI? Why or why not?

CHAPTER 21

Case study 2 (bacteria cause intestinal tract diseases)

Chapter outline

21.1 Introduction

Clostridium spp. are gram positive, straight to slightly curved bacilli, obligately anaerobic, spore-forming bacterial pathogen. The cultivation of anaerobic bacteria will be discussed in the other chapter. We will be focus on four major *Clostridium* spp., including *C. difficile*, *C. botulinum*, *C. tetani*, and *C. perfringens* and their related case studies.

21.2 *Clostridium difficile*

They are referred as "CDIFF" represents one of the most common nosocomial infections around the world and causes approximately three million cases of diarrhea and colitis per year in the U.S. They are recognized as the major causative agent of colitis (inflammation of the colon) and diarrhea that occur following 4–9 days extensive antibiotic intake in hospitals and chronic care facilities. The most common antibiotics associated with CDIFF infection are ampicillin, amoxicillin, cephalosporins, and clindamycin. One of the main symptoms of CDIFF is severe inflammation in the colonic tissue (mucosa) associated with destruction of cells of the colon (colonocytes) and cause plaque-like pseudo-membrane colitis accompanied with black water diarrhea. CDIFF colonize the gut and release toxin A and toxin B, which bind to certain receptors in the lining of the colon and

DOI: https://doi.org/10.1016/B978-0-12-821678-1.00026-5

ultimately cause diarrhea and inflammation of the large intestine, or colon (colitis). Therefore, the current diagnostic protocols include the serological testing the presence of A/B toxins. The treatments include (1) drinking lactic acid bacteria rich products such as yogurt, (2) fecal transplants using donated stool from other normal people, and (3) replace another antibiotic for example vancomycin.

21.3 *Clostridium botulinum*

They generate neurotoxins A, B, C, D, E, F, and G with toxins A, B, E are the most commonly related to neuroparalytic disease in the U.S. Typical neurological symptoms include double vision, difficulty speaking, difficulty swallowing, and peripheral muscle weakness (relaxing). The three types of botulism include (1) food botulism related to low acid (pH > 4.6, water activity ~0.85) home canned foods, (2) wound botulism involving infections in wounds contaminated with soil, and (3) infant botulism-"never feeding less than 1 year old newborns honey, otherwise will have floppiness." Equine antitoxins shipped from the airport is the primary treatment followed by antibiotics.

21.4 *Clostridium tetani*

Their typical symptom is referred as "lockjaw" (muscle contraction without relaxing), which occurred by the infections from a wound, burn, or ulcer. Their spores germinate and release tetanospasmin toxin. Death is usually caused by muscle spasms throughout the body interfering with muscles requiring for breathing. Neonatal tetanus usually occurred in developing countries with most home deliveries in unhygienic conditions. DTaP vaccine is required for kids in the U.S. and is an effective approach to control the infection from this pathogen.

21.5 *Clostridium perfringens*

C. perfringens is showing β-hemolytic double zones on blood agar and categorized into A, B, C, D, E five serotypes related to α-, β-, ε-, and l-toxins. They are most frequently responsible for released toxins and enzymes that cause tissue damage with typical symptom "Gas Gangrene". Gas gangrene always starts in a nearly anaerobic area such as wound or dead tissue area contaminated with clostridial spores. Their spores usually germinate to vegetative state and begin to multiply in the wound area. Their enzymes and toxins begin to spread outwards, damaging more tissue and producing gas by the enzymes attacking the tissues and splitting muscles apart to allow the infection to extend. The blood supply to muscle is terminated when the pressure of the gas rises above the arterial pressure and resulting in the absorption of toxins leads to septic shock. Gas gangrene can spread to a whole limb in a few hours following death.

C. perfringens is a food intoxication usually associated with contaminated meat such as hotpot cooking in a low temperature long time status. Their spore survives and starts to dominate in the low heat temperature conditions and toxins are released if entering the interties causing diarrhea, cramps, and abdominal pain. The control method includes thoroughly cooking foods containing poultry meat products with vegetables from soil and avoid the dangerous temperature zone of 40 to 140°F (4 to 60°C).

Urinary tract infection (UTI) has been discussed and practiced in previous lab sections. We will introduce two other pathogens related to UTI.

21.6 *Klebsiella pneumoniae*

They belong to the family of *Enterobacteriaceae*, are facultative gram negative, rods shape bacteria by the presence of a heavy capsular polysaccharide including 77 antigenic types. They may cause a severe pneumonia but only responsible for <1% of pneumonia cases requiring hospitalization in North America. They are most commonly associated with hospital acquired UTIs. The sputum samples from the patient are often referred to as "currant jelly" due to its appearance is often thick and blood tinged. Their colonies on MacConkey agars are large and highly mucoid and ferment lactose (pink colonies).

21.7 *Enterococcus*

They are the enterococci are responsible for nosocomial urinary tract infections. They are non-hemolytic or referred as γ-hemolytic on blood agars. They belong to the Group D streptococci. They can grow in high concentration of salt solution (up to 6.5% NaCl), resistant to bile and hydrolyze esculin to generate by-products esculetin, which reacting with ferric citrate producing black colonies.

21.8 Exercise 1

Please read the cases below and recognize the pathogens related to the case. You can discuss with your lab bench partner and record your results in your lab notebook and explain the reason why this pathogen is chosen for the case.

Case 1. On June 22, four people who attended a cookout fund raiser for the local Jaycees were hospitalized 24 hours later with difficulty swallowing, double vision, and slurred speech. Many patrons of the event recall some level of mild stomach upset in the late evening of the event. It is significant to note that many of the dishes were prepared by local individuals from home canned goods. Laboratory data from foods served at the event demonstrate gram-positive rods with subterminal endospores that grew anaerobically but not aerobically from a canned yellow tomatoes dish. What is the most probable etiological agent of this disease?

Case 2. A 60-year-old male was admitted to the hospital, 5 days after prostate surgery, complaining of dysuria. A qualitative and quantitative urine culture was obtained that revealed *Escherichia coli* in significant numbers. The physician placed the patient on a 7-day course of cephalexin. The urinary tract infection improved but the patient later experienced multiple watery loose stools, cramping, abdominal pain and dehydration. The physician ordered an A-B toxin test on a stool specimen, which returned positive. The patient was diagnosed with pseudomembranous colitis and was unsuccessfully treated with typical antibiotic therapy. A fecal transplant was performed, and the patient recovered in 24 to 48 h. What is the most probable etiological agent of this disease?

Case 3. A Gram-positive coccus was isolated from a 37-year-old female UTI. Laboratory tests indicate the organism grew on blood agar with no hemolytic zone, was catalase negative, and serologically belongs to Rebecca Lancefield group D. The microbiologist inoculated a bile esculin agar slant and incubated the tube at 37°C for 24 h showing black colonies.

CHAPTER 22

Isolation of phage and plaque forming units determination

Chapter outline

22.1 Warm-up questions

(1) What is a virus? (2) What is bacteriophage? (3) Describe the morphology of bacteriophage. (4) Describe the phage infection process. (5) Define lytic and lysogenic phages. Give examples of phages and their relationship to human diseases. (6) Describe a plaque. What is a phage titer? Please find the answers in your lecture slides or text book.

22.2 Objectives

Determine the titers (PFU/mL) of an *Escherichia coli* phage by practicing with 10-fold serial dilution technique.

22.3 Major materials

E. coli strain, prepared lambda phage, pipettes, pipette tips, dilution tubes, Kim-cap tubes, 0.1% buffered peptone water, bacterial agars

22.4 Introduction

Bacteriophages are viruses that infect only bacteria. The major phage structure includes nucleocapsid or head, and complex tail structure. The lytic life cycle of a bacteriophage includes: (1) Adsorption: Attachment sites on the bacteriophage tail adsorb to receptor sites on the cell wall of a susceptible host bacterium. (2) Penetration: A bacteriophage enzyme

Introductory Microbiology Lab Skills and Techniques in Food Science
DOI: https://doi.org/10.1016/B978-0-12-821678-1.00023-X

"drills" a hole in the bacterial cell wall and the bacteriophage injects its genome into the bacterium. This begins the eclipse period, the period in which no intact bacteriophages are seen within the bacterium. (3) Replication: Enzymes coded by the bacteriophage genome shut down the bacterium's macromolecular (protein, RNA, DNA) synthesis. The bacteriophage genome replicates and the bacterium's metabolic machinery is used to synthesize bacteriophage enzymes and bacteriophage structural components. (4) Maturation: The bacteriophage parts assemble around the genome. (5) Release: A bacteriophage-coded lysozyme breaks down the bacterial peptidoglycan causing osmotic lysis of the bacterium and release of the intact bacteriophages.

22.5 Lab practice procedures

Exercise 1. Preparing a phage dilution and plaque assay to determine PFU/ml: We will employ your serial dilution skills to determine the number of plaque forming units (pfu)/mL in a stock of a bacterial virus called lambda. The host for lambda (λ) is the bacterium *E. coli*. The whole procedure flowchart is shown in Fig. 22.1.

1. Make serial dilutions of the bacteriophage stock from 10^1 through 10^8 using microcentrifuge tubes and micropipettes. First draw out your dilution in your laboratory notebook using 20 mL of bacteriophage stock and 180 mL of nutrient broth in each of eight microcentrifuge tubes.
   a. Once your dilution has been approved by the instructor, set up a rack of eight sterile labeled microcentrifuge tubes and into each pipet 180 microliters of nutrient broth using the pipet and sterile pipette tips provided (nutrient broth is your phage titer diluent). Your instructor will demonstrate proper use of a micropipette. Please discard your pipette tips in the bio hazard sharps containers.
   b. With a new sterile pipette tip deliver 20 mL of phage stock solution to the 10^1 -microcentrifuge tube. Then using a new pipette tip for each successive dilution, you will continue until you complete the series.
2. Prepare the host cells for bacteriophage. (1) Set up four sterile 16 × 100 mm Kim-cap tubes labeled 10^4 through 10^8. These will be your "phage infection tubes." (2) Using a sterile 1 mL pipette dispenses 0.1 mL of an overnight culture of *E. coli* into each "phage infection tube." Dispose of the pipette into the bleach solution at your bench.

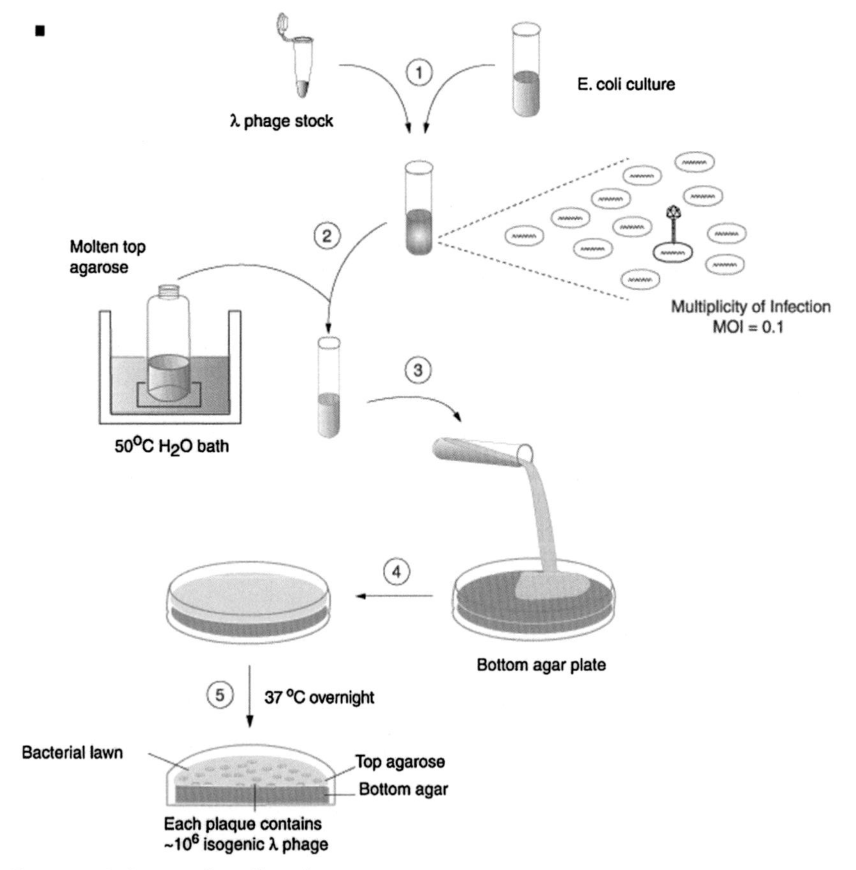

Fig. 22.1 ***Lab procedure flowchart.***

3. Label four agar plates each with the corresponding "phage infection tube" dilution 10^4 through 10^8. Note: plates must be prewarmed before use (the plates must be at room temperature or warmer).
4. Using your micropipette transfer 0.1 mL (100 mL) of the phage dilution into its corresponding phage infection tube (for example, transfer 100 mL of the 10^4 phage dilution into the 10^4 phage infection tube). After completing your phage dilution transfers mix all 4 phage infection tubes well by flicking.
5. Incubate your phage infection tubes at room temperature for 10 min. Do not disturb the tube during the incubation period.
6. Using a 5 mL plastic pipette, transfer 4 mL of melted agar into a phage infection tube (the melted top agar will be held at 50°C in the water bath). Mix well by pipetting up and down. Avoid creating bubbles (i.e., aerosols).

7. Transfer the mixture (melted top agar + phage + cells) onto the surface of the appropriate prewarmed agar plate. Gently swirl the mixture to cover the entire surface and avoid generating bubbles. Dispose of the pipette into the bleach solution at your bench.
8. Allow the top agar to solidify.
9. Incubate at 42°C for 24 h. The phage you are using grows at 42°C.
10. Return to lab and count the plaques. The plates should be covered with bacterial growth (this is called a "lawn"). Anywhere a successful infection has occurred, you should see a clear area (this is a plaque). Count the plaques on a plate that has between 20–200 plaques.
11. Estimate the number of PFN/mL in the original sample. This number is called the "titer". Your instructor will show you how to perform the calculation. PFN/mL = NC x RDF where NC is the number counted and RDF is the reciprocal of the dilution factor.
12. Record the titer in your lab notebook. Your instructor will check your results for accuracy.

*MOI, multiplicity of infection. The actual number of phages or viruses that will enter any given cell is a statistical process: some cells may absorb more than one virus particle while others may not absorb any. The probability that a cell will absorb virus particles when inoculated with an MOI of can be calculated for a given population using a Poisson distribution.

22.6 After class questions/readings

Isolation of *E. coli* phage from vegetable samples

Day 1

1. Obtain vegetable samples from local grocery stores.
2. Use *E. coli* ATCC13706, *E. coli* ATCC23631, and *E. coli* O157 ATCC700927 as phage propagation strains. Inoculate bacteria in 10 mL of tryptic soy broth (TSB).
3. Incubate bacterial culture at 37°C overnight.

Day 2

4. Mix 50 g of vegetables with 450 mL of buffered peptone water (BPW).
5. Take 50 ml of vegetable rinse and centrifuge at 2000 × *g* for 10 min.
6. Filter the supernatant with a 0.22 μm filter. Collect the filtrate in a fresh tube.

7. Treat the supernatant with chloroform at the ratio of 1:100 to kill any remaining bacteria. Mix thoroughly and let settle. (Hint: For 50 mL of supernatant, 0.5 mL of chloroform should be added.)
8. Add 5 mL of the supernatant and 0.1 mL of propagation strain to 20 mL of double-strength TSB. Incubate at 37°C overnight with 230 rpm shaking.

Day 3

9. Make 0.6% soft agar by adding 1.2 g of agar into 200 mL of TSB. Mix well and autoclave. Put soft agar in a 50°C water bath until ready for the soft agar overlay technique.
10. Centrifuge the culture from step 8 at 1000 × *g* for 10 min at 4°C.
11. Filter the supernatant through a 0.22 μm filter. Treat with chloroform (1:100). Mix thoroughly and let settle.
12. Dilute 0.1 mL of an overnight culture of propagation strain with 9.9 mL of TSB (1:100 dilution).
13. Mix 0.1 mL of the diluted culture with 4 mL of soft agar. Shake gently and immediately pour onto a tryptic soy agar (TSA) plate. Let soft agar solidify at room temperature.
14. Drop 10 μL of the filtrate from step 11 on TSA seeded with propagation strain. Let dry. Incubate the plate at 37°C overnight.

Day 4

15. Observe plaques as indications of bacteria lysis at the site of phage infection.

CHAPTER 23

Antibiotic susceptibility testing and evaluation of antiseptics/disinfectants

Chapter outline

23.1 Materials

Mueller-Hinton agar plates, Kirby-Beau test dispenser, antibiotic disks, filter paper disks, 0.9% saline solution, cotton swab

23.2 Introduction

Antibiotics are generated through metabolic results of one microorganism (molds/yeasts) to kill (~cidal) or inhibit (~static) another microorganism (bacteria). Antibiotics can be categorized as broad-spectrum or narrow-spectrum. A broad-spectrum antibiotic kills or inhibit both gram-positive and gram-negative bacteria. A narrow spectrum antibiotic will only affect either Gram-positive or Gram-negative bacterial strains. The early antibiotics were isolated from these natural sources; however, today many are genetically engineered to be even more effective than their natural counterparts. For an antibiotic to be useful to humans it must have the ability to destroy pathogens while being relatively non-toxic to the host organism. It should be chemically stable and be able to reach the part of the host organism in which the infection persists. The mechanism of antibiotics to against microorganism includes the following modes of cellular attack:

1. Cell wall synthesis inhibitors
2. Cell membrane inhibitors

Introductory Microbiology Lab Skills and Techniques in Food Science
DOI: https://doi.org/10.1016/B978-0-12-821678-1.00028-9

3. Protein synthesis inhibitors
4. Nucleic acid inhibitors
5. Competitive inhibitors

23.3 Exercise 1. The Kirby-Bauer disk diffusion assay

It is a semiquantitative test to compare how effective one antibiotic is to against another, or to measure the degree of antibiotic resistance in a bacterium. A pure culture of bacteria is first standardized using a McFarland standard 0.5 turbidity (approximately equal to 1 x 10^8 cells/mL). Then the pure culture is spread over the surface of a special medium, called Mueller-Hinton agar, to create a lawn of bacteria followed by using an antibiotic disk dispenser to drop the antibiotic disks onto the plate. After incubation, the antibiotic zone is measured to compare the standard table to determine the effect categorized as susceptible, resistant, or intermediate. The size of the zone of inhibition depends on various factors, including the rate of diffusion of a given drug in the medium, the degree of susceptibility of the organism to the drug, the number of organisms inoculated on the plate, and the growth rates.

23.3.1 Procedure

1. Collect strains of *S. aureus* (Gram +) and *E. coli* (Gram -) from the front bench.
2. Collect two Mueller-Hinton agar plates from the front bench.
3. Label your plate with the name of your organism, your name or initials and the date as shown in Fig. 23.1.

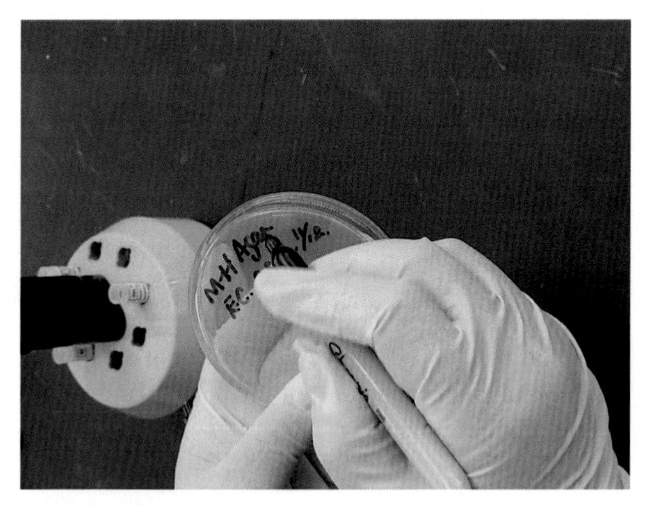

Fig. 23.1 ***Label the bottom of Mueller-Hinton agar.***

4. Use sterilized loop to pick colonies into the 0.9% saline solution and shake it as shown in Fig. 23.2.

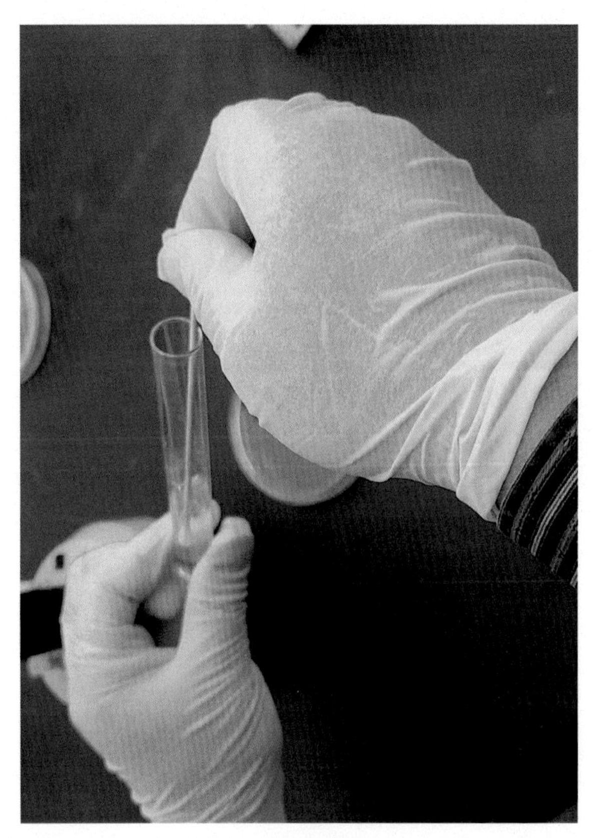

Fig. 23.2 ***Pick colonies into the 0.9% saline solution.***

5. Compare the turbidity to the McFarland Standard 0.5 Turbidity tube, and make minor adjustment if necessary.
6. Swab - inoculate the surface of your agar plate using a sterile cotton swab as shown in Fig. 23.3.
7. Dip the swab into your assigned broth culture and express the excess fluid from the swab by pressing it against the inside walls of the culture tube.
8. Swab the entire surface of the plate with a back and forth motion until the entire surface area is covered and inoculated. Rotate the plate 90° and repeat, but do not return to the assigned culture as shown in Fig. 23.4.

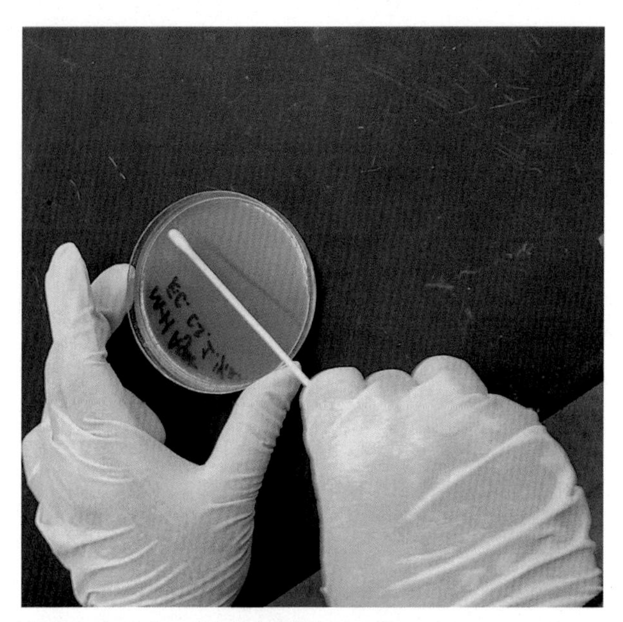

Fig. 23.3 ***Swab-inoculate the surface of the agar plate.***

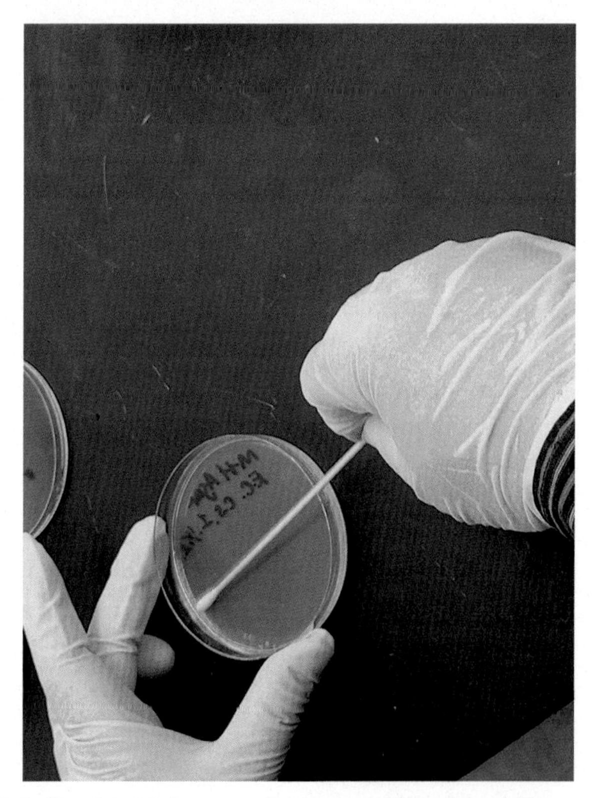

Fig. 23.4 ***Rotate the plate 90 degree and repeat swabbing inoculation.***

9. Allow the plate to dry (3–5 min).
10. Acquire an antibiotic disk dispenser to drop the antibiotic disks onto your plate. Remove the lid from the inoculated plate and place the appropriate dispenser over the plate in place of the lid (there are two dispensers, gram + and gram -). To dispense the antibiotics push down firmly on the plunger as shown in Fig. 23.5.

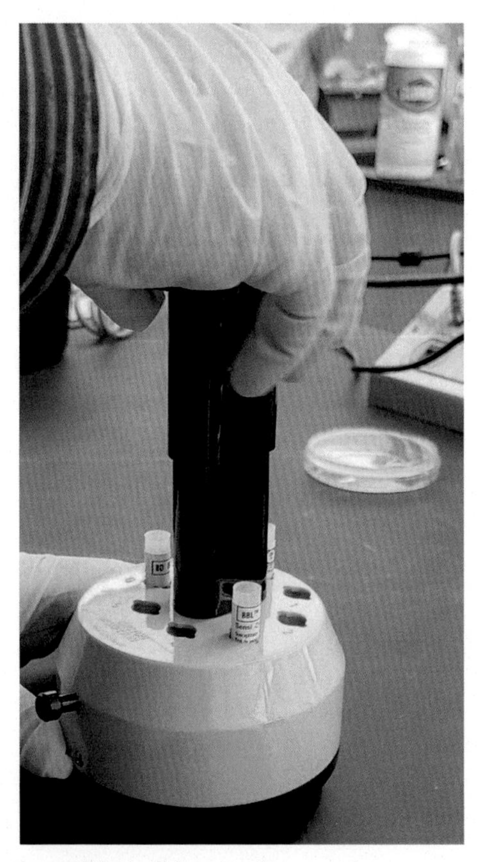

Fig. 23.5 ***Dispense the antibiotic disks onto the agar plate.***

11. Touch each disk with a flamed and sterilized pair of forceps to assure the disk does not fall when the plate is inverted for incubation (dip forceps into ethanol then flame). Tap each disk lightly onto the medium to secure it to the surface as shown in Fig. 23.6.
12. Invert the plates and incubate 16–18 h.
13. After incubation, measure the zone diameters with a metric ruler to the nearest whole millimeter as shown in Fig. 23.7.

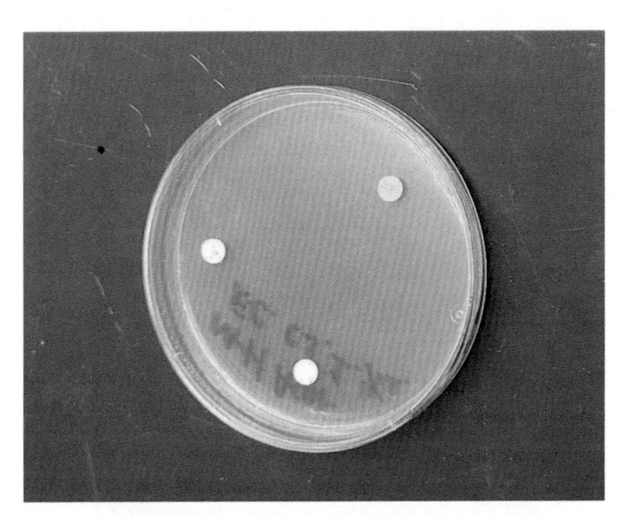

Fig. 23.6 ***Antibiotic disks attached on the agar plate.***

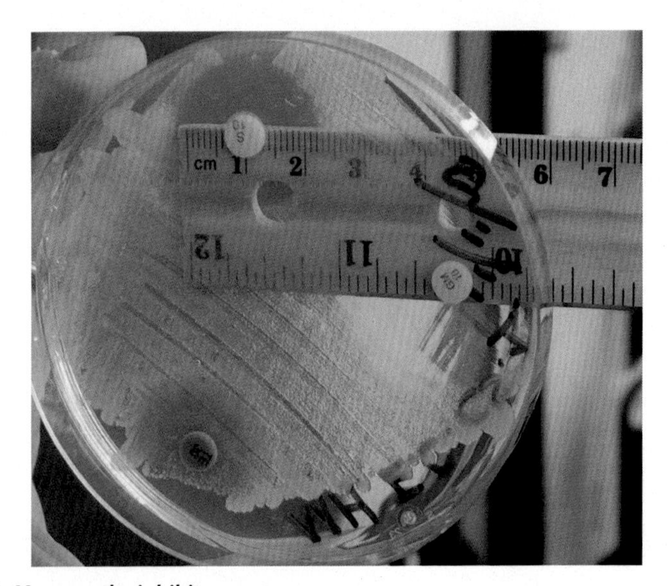

Fig. 23.7 ***Measure the inhibitory zone.***

14. Record the zone measurements for each antibiotic you have tested against your assigned species.

15. Compare a zone interpretation chart as shown in Table 23.1 to determine if your organism is susceptible (S), resistant (R), or intermediate (I). The zone of inhibition should vary with the type of organism.

16. Compare your result with your bench partners in the class.

Table 23.1 Results of antibiotic resistant tests.

Antibiotics	Disk code (concentration)	Diameter (inhibitory zone mm)			*Staphylococcus aureus*	*Escherichia coli*
		R	I	S	Zone diameter	Zone diameter
Ampicillin	10 mg-AM10	≤13	N/A	≥17		
Clindamycin	2 mg-CC2	≤14	15-20	≥21		
Erythromycin	15 mg-E15	≤13	14-22	≥23		
Gentamicin	10 mg-CN10	≤12	13-14	≥15		
Methicillin	5 mg-MET5	≤9	10-13	≥14		
Tetracycline	30 mg-TE30	≤14	15-18	≥19		
Vancomycin	10 mg-VA30	≤9	10-11	≥12		
Streptomycin	10 mg-S10	≤14	15-20	≥21		

23.4 Exercise 2. Evaluation of antimicrobials (filter paper disk method)

Filter paper discs, impregnated with standardized amounts of antimicrobials, are gently pressed on to the surface of the agar. The plates are incubated overnight while the antimicrobial diffuses from the disc into the agar. After incubation, the plates are examined for the presence of zones of inhibition (clear rings around the antibiotic disc). If there is no inhibition, growth extends up to the rim of the disks on all sides, then the organism is reported as resistant (R) to the antimicrobial agent. If a zone of inhibition surrounds the disk, the organism is not automatically considered susceptible (S) to the antimicrobials.

23.4.1 Procedure

1. Obtain your assigned cultures of *S. aureus* (Gram +), *E. coli* (Gram -), *Pseudomonas aeruginosa* (Gram -), or *Corynebacterium xerosis* (Gram +) from the front desk.
2. Obtain one nutrient agar plate from the front desk.
3. Label the bottom of the plate with your organism, your name and date. You will divide the plate into three sections and label each with one of three antimicrobials being tested.
4. Swab the organism to be tested onto the surface of a nutrient agar plate using the same procedure as outlined for the Kirby-Bauer method as exercise 1.
5. Use your flamed and sterilized forceps pick up a sterile filter paper disks and dip them halfway into the phenol, formaldehyde or iodine solution.
6. Place the disks in the center of the *appropriate* section of the inoculated plate and gently press it to the agar surface.

Table 23.2 Results of antimicrobial disk tests.

Antimicrobials	***Staphylococcus aureus*** **inhibitory zone (mm)**	***Escherichia coli*** **inhibitory zone (mm)**
Phenol		
Formaldehyde		
Iodine		

7. Repeat this for the other *antimicrobials*, placing each respective disk in one of the three labeled sections.
8. Invert plates and incubate at 37°C.
9. Check plates at 24 h and measure the zone of inhibition in millimeters.
10. Return after another 24 h and repeat the measurements and check to see whether the measurements change.
11. This is a comparative test. You will evaluate the efficacy of a particular antimicrobial on a particular organism by comparing the millimeters of inhibition. There is no standardized chart for this.
12. Fill the results in the Table 23.2.

23.5 After class questions/readings

1. Discuss antibiotics, antimicrobials, zone of inhibition, Kirby-Bauer method (a standardized method). What does it mean to be standardized?
2. What factors can influence the size of the zone?
3. Discuss different types of antimicrobials.
4. Define antiseptics, disinfectants, sterilants, bacteriocidal, bacteriostatic.
5. Discuss mechanism of action of phenol, formaldehyde and iodine.
6. Minimal inhibitory concentration (MIC) is the lowest concentration of antibiotic that inhibits the growth of specific amount of bacteria. MIC can be used to compare the level of antibiotic resistance between bacteria. The higher the MIC, the less susceptible the bacteria to the antibiotic. MIC can be obtained by broth dilution/microdilution test. Bacteria are then interpreted as "susceptible," "resistant," or "intermediate resistant" according to the resistance breakpoint published by the Clinical and Laboratory Standard Institute (CLSI).
7. Antibiotic resistance mechanisms include antibiotic inactivation, target modification, and efflux pumps. These mechanisms are mediated by various antibiotic resistance genes. While Kirby-Bauer disk diffusion assay measures the phenotypic susceptibility of bacteria to antibiotics, polymerase chain reaction (PCR) can be used to detect corresponding antibiotic resistance genes using gene-specific primers.

CHAPTER 24
Microbial counts of food product

Chapter outline

24.1 Materials

Nutrient agar plates, 0.1% buffered peptone water, 15 mL dilution tubes, pipettes, pipette tips, ground beef, frankfurters, tomatoes, food sample bags

24.2 Introduction

Each food product will contain a different number of organisms depending on the origin of the sample and the processing that was implemented by the food industry. Many food products that we believe to be "sterile" when packaged will contain normally occurring bacteria from the environment or the host/source of that product (i.e., dairy, beef, and vegetables). Some food-borne pathogens are associated with outbreaks in related food products. For example, ground beef could be contaminated with *Escherichia coli* O157:H7 and United States Department of Agriculture, Food Safety and Inspection service required a "zero-tolerance." Frankfurters could be contaminated with Listeria monocytogenes and U.S. Food and Drug Administration established "three-alternatives" to control it on ready-to-eat meat products. Tomatoes, especially the stem-scar parts, can be easily penetrated by pathogens such as *Salmonella* spp. and *E. coli* O157:H7. U.S.-FDA published the guidelines of temperature requirement of tomato washing waters.

24.3 Exercise 1. Bacterial count on food, spread plate technique

Students will work in groups of 2. Your group will be assigned 1 food product to quantitate $\log_{10}$CFU/g or $\log_{10}$CFU/cm^2.

Introductory Microbiology Lab Skills and Techniques in Food Science
DOI: https://doi.org/10.1016/B978-0-12-821678-1.00025-3

Each group will be given an assigned food product, boxes of 1000 μl and 100 μl pipets, and three nutrient agars.

24.4 Procedures for preparing food samples before analyzing

For ground beef: Using aseptic technique, weigh out 25 grams of your assigned food onto sterile weighing paper (the instructor will demonstrate how to use a pan balance). Add the 25 grams of food plus 100 mL of buffered peptone water to a sterile food sampling bag and place the bag into stomacher for 1 min as shown in Fig. 24.1.

Fig. 24.1 ***Place the bag into stomacher.***

For frankfurters: Using aseptic technique, add one frankfurter plus 100 mL of buffered peptone water into a sterile food sampling bag as shown in Fig. 24.2, rolling the bag as shown in Fig. 24.3, and shake 30 s as shown in Fig. 24.4.

For tomatoes: Using aseptic technique, add one tomato plus 100 mL of buffered peptone water into a sterile food sampling bag, shake 30 s, message 30 s, and shake again 30 s.

Below you will see a chart that recommends a dilution scheme for your assigned food product as shown in Table 24.1. First you will draw out your specific dilution in your laboratory notebook. Once you have the dilution on paper you will label your three sterile petri dishes, then using

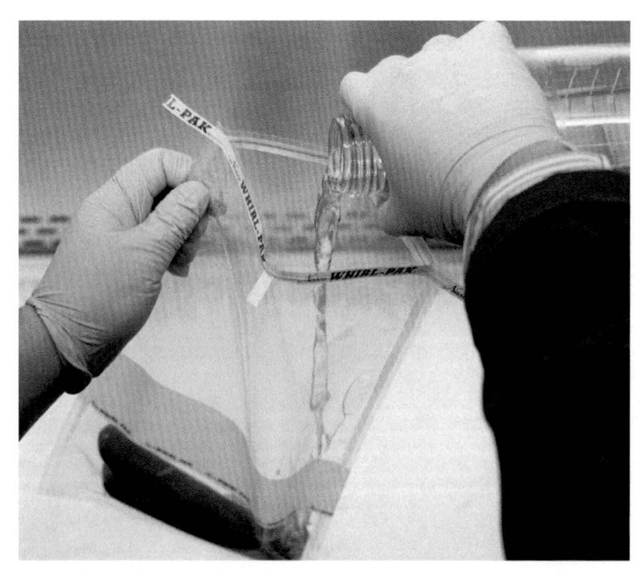

Fig. 24.2 ***Add 100 mL nutrient solution into food sample bag.***

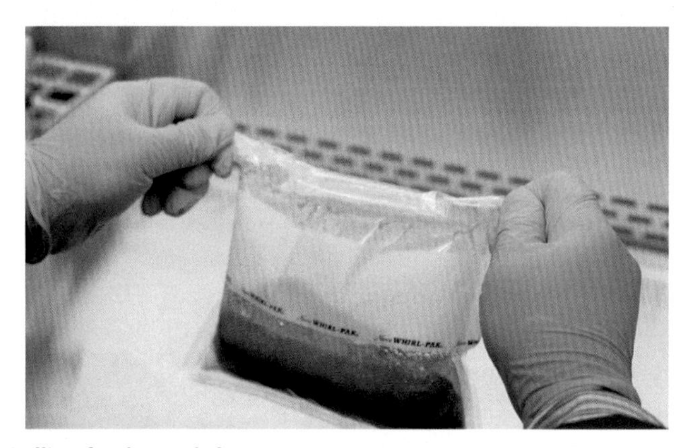

Fig. 24.3 ***Rolling food sample bag.***

your 1000 μL and 100 μL pipets to set up your dilution. You may ask the instructor to verify your dilution scheme before you begin.

1. Locate your food product below and draw your respective dilution scheme in your laboratory notebook.
2. Use 1000 μL and 100 μL pipets to make 10-fold or 100-fold serial dilutions in 9.9 mL or 9.0 dilution tubes (0.1% buffered peptone water) based on your dilution scheme.
3. Shake the diluted tube 25 times (your instructor will demonstrate the proper technique).

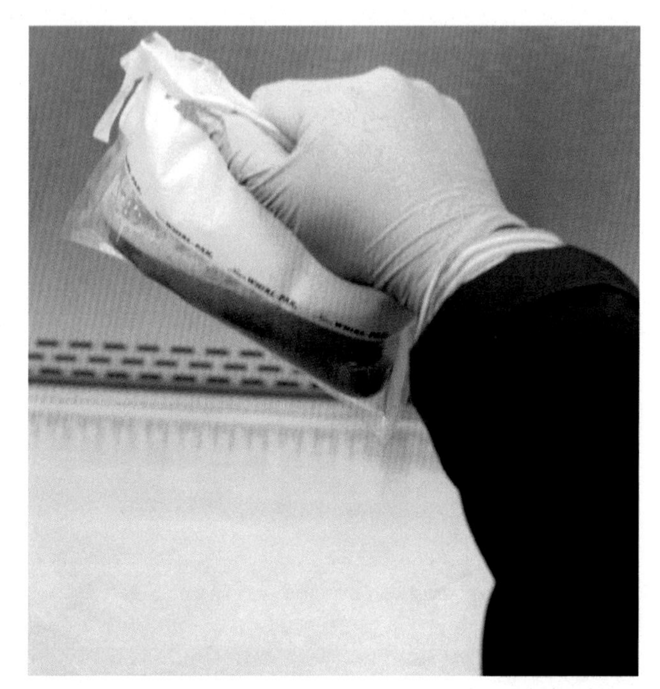

Fig. 24.4 ***Shake food sample bag.***

4. Spread plating 0.1 mL diluted solution onto nutrient agars.
5. After the plates have dried, incubate at 37°C for 24 h.
6. Count the colonies on the plate that has between 30 and 300 colonies.
7. For ground beef and tomatoes, calculating $\log_{10}$ CFU/g = $\log_{10}${[final colony numbers (30–300) × dilution factor] × (100 + food weight gram)/food weight gram}.
8. For frankfurters, calculating $\log_{10}$ CFU/g = $\log_{10}${[final colony numbers (30–300) × dilution factor] × 100/surface area = 22 cm^2}.

Table 24.1 Dilution scheme of food samples.

Assigned food products and their recommended dilutions	
Ground beef	10^1, 10^3, and 10^5
Tomato	10^1, 10^3, and 10^4
Frankfurter (hotdog)	10^1, 10^2, and 10^3

24.5 After class questions

1. Please complete the flowchart of preparing tomato samples.
2. Sketch out a dilution scheme using 99 mL or 9 mL dilution blanks (or both) that would produce a countable plate for an original stock culture with 38,000,000 CFU/mL (Hint: we add 0.1 mL onto agar plates for spread plating).
3. If an original bacteria culture was diluted 10^5 times using serial dilution and 0.1 mL of the diluted culture was spread on an agar plate that produced 172 colonies after incubation, what was the concentration of the original bacteria culture?
4. Quantitative PCR (qPCR) can be used to detect and quantify bacteria in food samples. Unlike culturing methods that rely on the multiplication of target bacteria during pre-enrichment and selective enrichment, qPCR detects bacterial DNA in the food sample and therefore bypasses the long incubation time. However, the presence of DNA alone does not indicate the viability of bacterial cells. Based on what is presented above, what are the major advantages and disadvantages of qPCR compared to culturing methods? How would you overcome the disadvantages of qPCR in bacteria detection?

CHAPTER 25

Total plate counts & coliform counts of pond water

Chapter outline

25.1 Materials

Membrane filter unit, Pond water, Petri films, 9.9 mL and 9.0 mL 0.1% buffered peptone water, pipette tips, pipettes

25.2 Introduction

Water utilities ensure the safety of our drinking water by conducting standardized testing procedures to detect the presence of potential disease-causing organisms. These pathogens enter our water supply directly by human or animal feces, or indirectly through improperly treated sewage. The organisms of most concern are the intestinal pathogens, particularly those that cause typhoid (enteric) fever and bacillary dysentery (colitis). Since it is impossible to test each water sample for every pathogen. it is much more cost effective to test for the presence of nonpathogenic intestinal organisms such as *Escherichia coli* and other related species as indicators. *E. coli* is a normal inhabitant of the intestinal tract and is not normally found in fresh water. If the presence of *E. coli* was detected in water samples then we assume there has been fecal contamination of that water supply and potentially presence of intestinal pathogens. A group of organisms known as Coliform Bacteria are a commonly used as indicators of sanitary quality in foods and water.

DOI: https://doi.org/10.1016/B978-0-12-821678-1.00003-4

25.3 Coliform vs fecal coliform

Coliform bacteria are defined as rod-shaped gram-negative nonendospore forming bacteria that ferment lactose with the production of acid and gas at 35–37°C. Fecal coliforms, the vast majority being *E. coli*, are considered to be more directly associated with fecal contamination from warm blooded vertebrates than are other members of the coliforms. Because coliform counts are inadequate to differentiate between fecal and nonfecal contamination, fecal coliform testing was developed. Fecal coliforms are coliforms that ferment lactose in EC medium (EC stands for *E. coli*) with gas production within 48 h at 44.5°C (not 35–37°C for coliform testing). The U.S.-EPA regulation require both zero limit for both coliform and fecal coliform in drinking water.

25.4 Exercise 1. The membrane filter method

In the membrane filter technique 100 mL or greater of a test sample is passed through a filter with pores small enough to retain all bacteria in the sample. The filter is then placed on a selective medium that allows for the detection of coliforms.

25.4.1 Procedure

1. Assemble the membrane filter unit as shown in Fig. 25.1.
   a. Unwrap the sterilized millipore filtration unit including funnel, cover, filter holder as shown in Fig. 25.1.

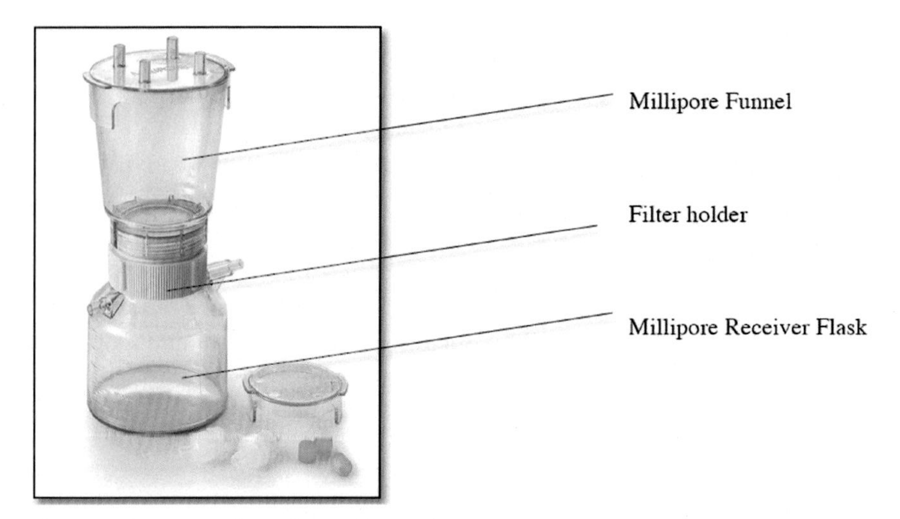

Fig. 25.1 ***Structure of membrane filter unit.***

  b. Place the sterile assembled millipore funnel onto a receiver flask.
  c. Unscrew the funnel from the filter holder.
  d. Aseptically unwrap and insert a filter into the filter holder using sterile forceps (be sure you remove and discard the wax paper from beneath the filter and place the filter, grid up, onto the holder).
  e. Assure there is an O-ring seal in the funnel where it meets the filter then screw the funnel onto the filter holder.

2. Pour 100 mL of water sample into the millipore funnel unit.
3. Attach the vacuum hose and suction the sample through the filter.
4. Carefully disassemble the filtration unit.
5. Use ethanol to flame sterilize your forceps an aseptically remove the filter.
6. Place the membrane filter (grid side up, 60 mm) onto the surface of an m-Endo agar plate.
7. Incubate plate at 37°C for 24 h.
8. Count only those colonies that produce a green metallic sheen.
9. Repeat this procedure with the other plates prepared by your classmates.
10. Record your results in your notebook.

25.5 Exercise 2. The petri-film method

Petri-film plate is a "paper medium" and created by food safety division of 3M Corporation. They contain a cold-water-soluble gelling agent with dehydrated nutrients. A foam barrier accommodating 1.0 mL of solutions in the 20 cm^2 circular area of the plating surface is covered by a top film. There are mainly three different petri films commercially available for testing population of total aerobic bacterial counts, generic *E. coli* and coliforms, and yeast and molds.

25.5.1 Procedure

1. Obtaining a bottle of pond water from the front desk.
2. Draw your respective dilution scheme in your laboratory notebook as shown in Table 25.1. The final dilution is the same as tube dilution factor due to the 1 mL of solution added onto the petri films.
3. Label each petri film of APCs and *E. coli*/coliforms.
4. Use 1000 μL and 100 μl pipets to make 10-fold or 100-fold serial dilutions in 9.9 mL or 9.0 dilution tubes (0.1% buffered peptone water) based on your dilution scheme as shown in Table 25.1.

Table 25.1 Dilution scheme of pond water.

Recommended dilutions of pond water	
Aerobic plate counts	10^1, 10^3, and 10^5
Escherichia coli and total coliforms	10^0, 10^1, and 10^2

5. Shake the diluted tube 10 times.
6. Add 1.0 mL of diluted solution onto the petri film as shown in Figs. 25.2 and 25.3.

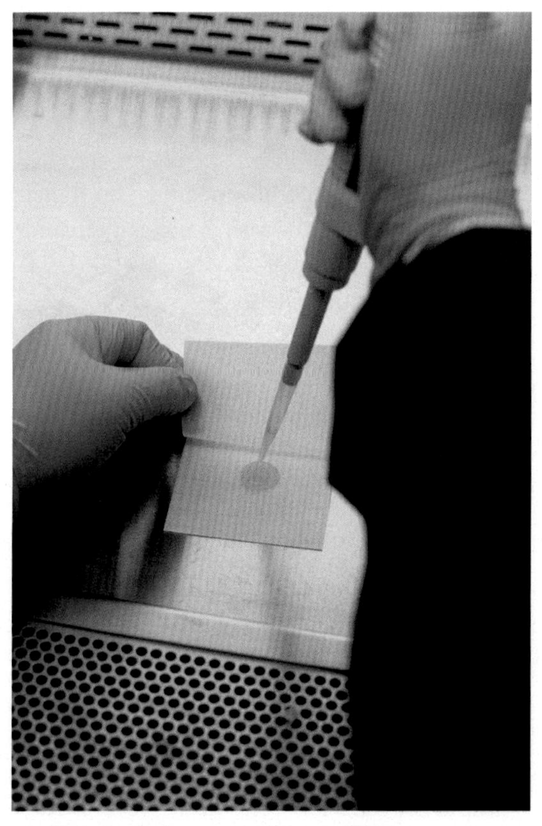

Fig. 25.2 ***Add 1 mL diluted solution onto APCs petri film.***

7. For APCs, cover the plating area with film and pressed with plastic presser as shown in Fig. 25.4.
8. For *E. coli*/coliforms, cover the plating area with film only without pressing.

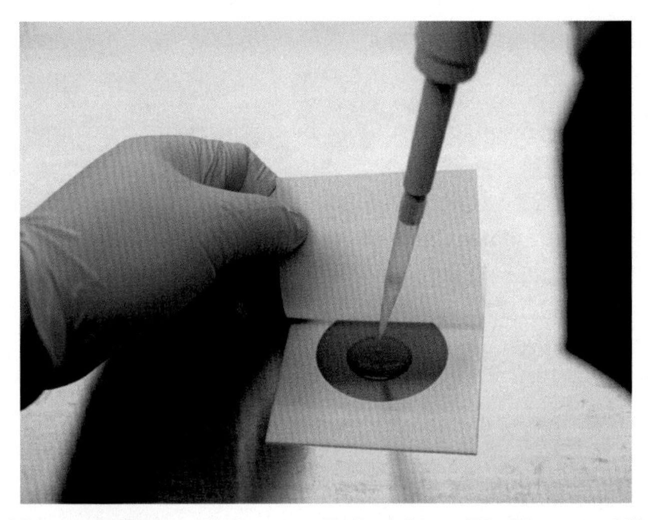

Fig. 25.3 ***Add 1.0 mL diluted solution onto*** **Escherichia coli*/coliform petri film.***

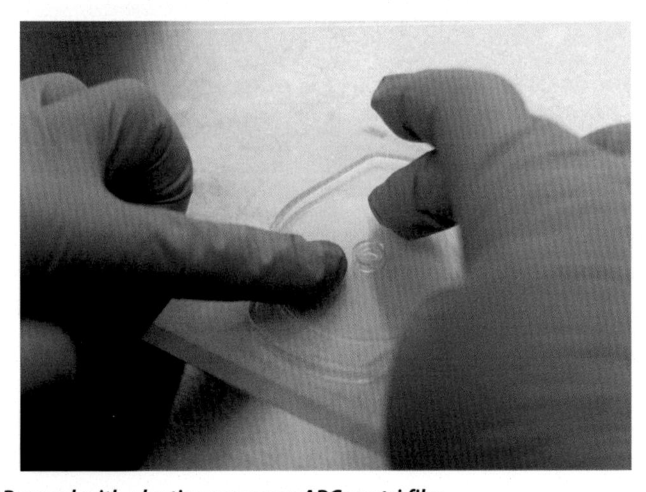

Fig. 25.4 ***Pressed with plastic presser on APCs petri film.***

9. After the plates have dried, incubate at 37°C for 24–48 h.
10. Count the colonies on the petri film that has between 30 and 300 colonies.
11. Colonies of pink color indicates coliforms, colonies of blue color with gas bubble indicates generic *E. coli*.
12. Total coliform counts = counts of coliform (pink color) + counts of generic E. coli (blue colonies with gas bubbles).
13. Calculating $\log_{10}$ CFU/g = $\log_{10}$ final colony numbers (30–300) × dilution factor.

25.6 After class questions

1. Compare coliform vs fecal coliform?
2. Describe the advantages of membrane filter method and petri-film method.
3. Results of Fig. 25.5 is from a previous student, please label *Escherichia coli* and APCs and calculate the concentration of total coliform and APCs in this pond water sample.
4. Describe the U.S. EPA regulation of drinking water related to microbial safety and quality.

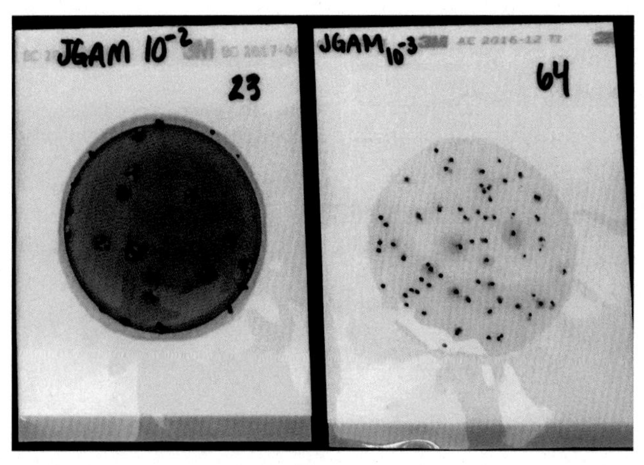

Fig. 25.5 ***Results of pond water on petri film.***

Appendices

General microbiology laboratory (AEM 341)

Fall 2020 course information and syllabus (sample)

Instructor:	Dr. Cangliang Shen
Office:	2423 New Agriculture Science Building
Telephone:	304-293-2691 (Office, please leave message) 970-222-2975 (Cell Phone, please leave message)
Email:	cashen@mail.wvu.edu
Office hours:	Tuesday and Thursday afternoon 10:00–10:45 or by appointment
Home page	http://nutrition.wvu.edu/faculty_staff/cangliang-shen
Lecture hours:	Tuesday 12:30–1:45 pm, 2:00–3:15 pm, or 3:30–4:45 pm
Lecture room:	SAS-E 1011
Recommend Text Book	Cangliang Shen and Yifan Zhang. Introductory Microbiology Lab Skill and Techniques in Food Science. 2021. Elsevier Inc. ISBN 9780128216781.

Date	Lab Work
	Seating, Laboratory Safety Rules, Equipment Assignments
Sep 01	Ex. 1 – Bright-field Microscopy (discussion and observation microorganism in prepared slides)
	Ex. 2 – Preparation of Smears and Simple Stain
Sep 08	Ex. 3 – Gram Stain
	Ex. 4 – Endospore Stain
15	Ex. 5 – Aseptic Technique-I Bacteria Transfer (plate to slant, plate to broth, streak-plating)
	Ex. 6 – Nutritional Requirements (Synthetic and Complex Media)
	Ex. 7 – Selective & Differential Media (McConkey, Manitol Salt, Blood agar)
22	Ex. 8 – Pure Culture Techniques- Spread plating
	Ex. 9 – Dilution technique (Discussion)
29	Pure Culture Techniques- Pour plating
	Ex. 10 – Temperature Requirement
	Ex. 11 – Oxygen Requirements
	Ex. 12 – Cultivation of Anaerobes (Demonstration)
	Mid-term exam review (Q & A section)
Oct. 06	**Mid-term exam (paper only, 60 points)**
Oct. 13	Ex. 13 – Urease Test
	Ex. 14 – Carbohydrate Fermentation
	Ex. 15 – Catalase Test

(*continued*)

DOI: https://doi.org/10.1016/B978-0-12-821678-1.00022-8

Date	Lab Work
	Ex. 16 – Nitrate Reduction
	Ex. 17 – Decarboxylation & Deamination of Amino Acids
20	Ex. 18 – IMViC Test
	Ex. 19 – Hydrogen Sulfide Test
	Ex. 20 – Oxidase Test
	Ex. 21 – Extracellular Enzymatic Activities of Microorganisms
27	Ex. 22 – Urine test (UTI discussion)
	Ex. 23 – Osmotic Pressure and Bacterial Growth
Nov 10	Ex. 24 – Antibiotic Sensitivity Testing
	Ex. 25 – Total Plate Count & Coliform Count of Pond Water
Nov. 17	Ex. 28 Microbial Counts of Food Product
	Final exam review
24	**FINAL LABORATORY EXAMINATION (paper only, 100 points)**
	Laboratory Check-Out
	Return Lab Exams & All Semester Scores

Attendance requirements

Laboratory attendance is mandatory since absence records will be used in determination of grades. **For each unexcused absence, 1 point will be subtracted from the laboratory grade.** You must contact me no later than the laboratory period immediately following the date of your absence in order to determine whether your absence will be excused without loss of 2 points. ***Missed labs cannot be made up later.***

Grading

See lecture syllabus for grading system.

Laboratory:

Dilution test	15
Attendance	25
Lab notebook	(20)
Mid-term exam	60
Final Exam	100
Total	200 points

Handicaps/Limitations/Concerns

If you are a person with a disability/limitation/concern and anticipate needing any type of accommodation in order to participate in this class, please contact Disability Services (293-6700) to determine what action is appropriate.

Please note: If you suspect that you are immunocompromised (diabetic, immune deficiency disease, pregnant, etc.), *it is your responsibility to contact your physician to determine whether additional safety precautions* are needed in order for you to safely complete the laboratory requirement for this course. Please also let Dr. Shen know, we will avoid you doing certain lab assignments.

Laboratory safety guidelines

It is each student's responsibility to follow oral directions of the laboratory instructor regarding safe operating procedures in the laboratory. In addition, each student **must carefully read and be thoroughly familiar with** the written safety rules/procedures accompanying this syllabus.

Lab notebook record

Notebooks can be checked after each lab section before you leave the classroom.

The front cover of your notebook will be labeled with following information: your name, course name, professor's name, and semester.

Reserve the first three pages of your notebook for a table of contents. You will update the table of contents each week.

Number all pages.

Each notebook entry for any given lab period will be formatted in the following way:

1. The Title.
2. A Purpose statement: Describe the purpose of the exercise. Ask yourself "What are the goals of this exercise?"
3. A Materials and Methods section: The Materials and Methods for each lab will described at the beginning of the period by the instructor. You will have an opportunity to copy this information into your notebook at the beginning of each lab period. You should describe what was needed and the steps taken (including any modifications that were made). Be sure to use CORRECT SPELLING for all microorganism names and italicize scientific names. Points will be deducted for incorrectly formatted scientific names. NOTE: Copying from the lab manual is illegal.
4. A Results section: Record all observations in your lab notebook. Colored pencils/pens should be used to illustrate results (i.e. observations made with the microscope) - all figures/tables must have a title and legend (a description of what is being shown - label all relevant information)

5. A Discussion - Summarize your findings and discuss how the exercise helped you understand the learning objectives. Describe why something may not have worked, and what you would do differently next time to improve the outcome.
   Each section must be LABELED.
   Points will be deducted for incomplete entries and sloppiness.

COVID-19 syllabus statement

WVU is committed to maintaining a safe learning environment for all students, faculty, and staff. Should campus operations change because of health concerns related to the COVID-19 pandemic, it is possible that this course will move to a fully online delivery format. If that occurs, students will be advised of technical and/or equipment requirements, including remote proctoring software.

In a face-to-face environment, our commitment to safety requires students, staff, and instructors to observe the social distancing and personal protective equipment (PPE) guidelines always set by the University. While in class, students will sit in assigned seats when applicable and wear the required PPE. Should a student forget to bring the required PPE, PPE will be available in the building for students to acquire. Students who fail to comply will be dismissed from the classroom for the class period and may be referred to the Office of Student Conduct for further sanctions.

If a student becomes sick or is required to quarantine during the semester, they should notify the instructor. The student should work with the instructor to develop a plan to receive the necessary course content, activities, and assessments to complete the course learning outcomes.

Example of Students Lab Notebook Record-1

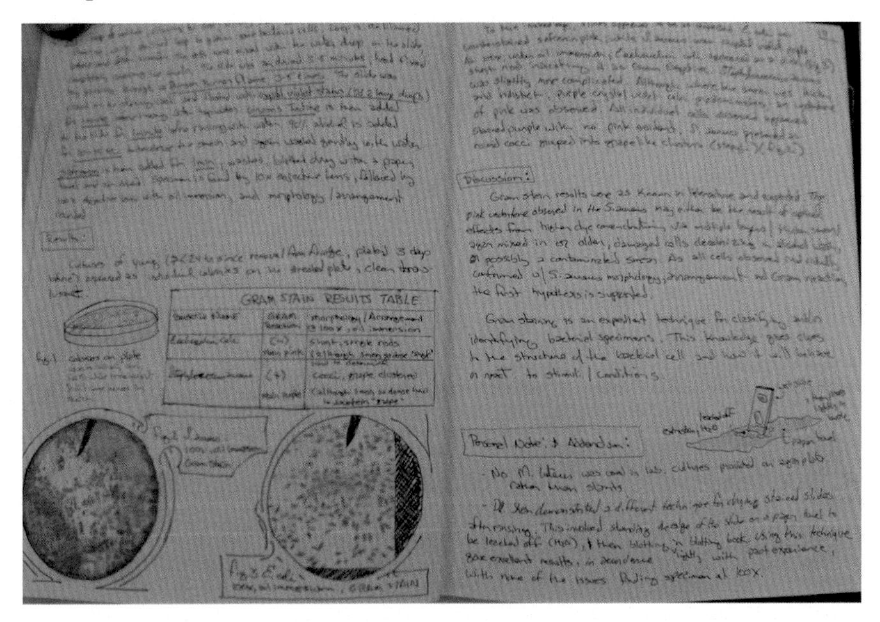

Evaluation: This student did excellent job of recording all lab details with very good hand-drawing pictures, however the handwriting needs to be slowing down little bit and write more clearly.

Example of Students Lab Notebook Record-2

*Results:

Bacterial Growth Results:

	Escherichia coli	Pseudomonas aeruginosa	Staphylococcus aureus
MacConkey Agar:	Grew properly	Grew properly	Doesn't grow in this medium
Mannitol Salt Agar:	Doesn't grow in this medium	Doesn't grow in this medium	Grew properly
Blood Agar:	Gamma hemolytic (No hemolysis, degradation of red blood cells)	Light β-hemolysis (moderate degradation of red blood cells around bacterial streak)	β-hemolysis (Extensive degrad... of Red Blood cel... within proximit... of bacterial str...

MacConkey Agar   Blood Agar   Mannitol Salt   TSA (nutrient) Agar

*Discussion:

In this lab, students used their looping and bacterial transfer/streaking knowledge to prepare and observe four different types of agar plates with the same three bacterial species: E. coli (EC), Pseudomonas aeruginosa (PA), and Staphylococcus aureus (SA). The four agar plates tested included that of MacConkey agar, blood agar, Mannitol Salt, as well as TSA nutrient agar. Within the MacConkey agar plate,

Evaluation: This student did excellent job of recording lab details with very good handwriting, especially including printed pictures of lab results and very detail lab results table and discussion.

Index

Page numbers followed by "*f*" and "*t*" indicate, figures and tables respectively.

Printed in the United States
by Baker & Taylor Publisher Services